SEEING
COLOUR

A Journey Through
Goethe's World of Colour

First published by Floris Books in 2022
Some parts were published in *Experience Colour*,
Ruskin Mill Trust, 2018

 Also available as an eBook

British Library CIP Data available
ISBN 978-178250-780-2
Printed in Poland through Hussar

 Floris Books supports sustainable forest management by
printing this book on materials made from wood that
comes from responsible sources and reclaimed material

MIX
Paper from
responsible sources
FSC® C167221
FSC
www.fsc.org

SEEING COLOUR

A Journey Through Goethe's World of Colour

Nora Löbe
Matthias Rang
Troy Vine

Floris Books

Contents

Foreword

When the German poet Goethe returned to Weimar from his travels in Italy, he resolved to discover the truth with regards to colour, especially the aesthetic use of colour. In Italy Goethe had painted and conversed with artists about their work, but he sought a deeper understanding of colour than they could provide. Once home he consulted science books, which were of little help. They spoke of Newton's corpuscles or Huygens's waves – models of light and colour that were of little use to those who sought the basis for the artistic use of colour. On the verge of abandoning his project, Goethe turned away from abstract models and towards the phenomena of colour themselves. If he could not find a Farbenlehre adequate to his needs, he was confident that he could undertake the research and study necessary to write one himself. The result was the publication in 1810 of Goethe's comprehensive *Farbenlehre*, or *Theory of Colour*, in three volumes: Historical Part, Polemical Part, and Didactic Part. The journey taken by Goethe through the world of light, darkness and colour is one we can all still take; this volume marks out a path for such a journey.

Nora Löbe, Matthias Rang and Troy Vine are expert guides, leading us from the first faltering steps of Newton and Goethe, to their contrasting ways of seeing colour. Step-by-step they guide us through increasingly complex colour phenomena, starting with those colours that arise from the eye itself and culminating in the phenomena of additive and subtractive colour mixing. The authors are careful to respect Goethe's method of working, which stays close to the phenomena themselves and resists all conventional efforts at explaining colour effects by underlying mechanisms. Instead of hypothetical mechanisms, Goethe sought for what he termed the archetypal phenomena of colour, which he found in the blue of the sky and the red of the sunset.

> The ultimate goal would be: to grasp that everything in the realm of fact is already theory. The blue of the sky shows us the basic law of chromatics. Let us not seek for something behind the phenomena – they themselves are the theory.

The phenomena 'themselves are the theory'. Within these few words Goethe secreted a whole philosophy of knowledge. We are not surprised, therefore, by Wittgenstein's appreciation of Goethe's phenomenological method and its implications for a logic of colour. The English painter J.M.W. Turner became interested in Goethe's approach, so much so that he painted two studies of the archetypal phenomena of Goethe.

Goethe's emphasis on colour experience reflected his confidence that a true understanding of colour can only occur once we possess the requisite capacities. In this way Goethe's science was deeply concerned with education as formation. We learn to see colours; indeed, we learn to see nearly everything. As Goethe wrote: 'Every object, well-contemplated, opens a new organ in us.' In this sense education is concerned with the formation of the human being.

Löbe, Rang and Vine invite us on a journey into a world of colour experience. In joining them we undertake the important project of shaping new capacities by which we can know our world more fully.

Arthur Zajonc
Emeritus Professor of Physics, Amherst College

Introduction

During revolutions scientists see new and different things when looking with familiar instruments in places they have looked before. It is rather as if the professional community had been suddenly transported to another planet where familiar objects are seen in a different light and are joined by unfamiliar ones as well.

The Structure of Scientific Revolutions
Thomas Kuhn

A journey through a foreign land can forever change how we see the world. It can extend our horizons, open up new perspectives, and when we return it can be our own familiar land that seems the most foreign of all. We see things we had not seen before, and even familiar things we now see in a new and unfamiliar light. Such a journey has transformed us, and as it transforms us, it transforms our world too. This book is an invitation to go on such a journey. The journey in question is through Johann Wolfgang von Goethe's world of colour.

Before setting out we will briefly consider Goethe's own journey, which he describes in the concluding chapter of his three-part monumental work on colour, *Zur Farbenlehre* (towards a teaching on colour), published in 1810. Goethe became interested in colour on his Italian journey in 1786–88. He wanted to learn about painting and observed how painters were able to speak knowledgeably about all aspects of their practice except colour. He therefore decided to discover the fundamental principles of colour for himself.

Goethe recognised that one first needs to approach colour as a natural phenomenon if one wants to learn anything useful for art. So, on his return to Weimar, he set about repeating Isaac Newton's famous prism experiments, which had laid the foundation for optics.

Convinced, like everyone else, that all the colours are contained in sunlight, he borrowed some prisms, found a suitable room he could darken, and made a small hole in the window shutter. However, other activities prevented him from repeating Newton's experiments and eventually the owner of the prims demanded their return.

Goethe described what happened next as follows:

> I had already taken the box to give to the messenger when it occurred to me that I should at least take a brief look through a prism; something which I hadn't done since my earliest youth. I remembered well that everything appeared coloured through the prism, but in what manner I had forgotten. At that moment I was standing in a room painted completely white. As I put the prism to my eye I expected, with Newton's theory in mind, to see the entire white wall covered in bands of different colour; to see the light that was reflected from the wall into the eye split into many coloured lights.
>
> But how astonished was I when the white wall seen through the prism remained white just as before. Only where something dark bordered it did a more or less determinate colour appear. The window-bars appeared most vividly coloured, whereas not the slightest trace of colour could be seen on the light-grey sky outside. I did not have to deliberate long to realise that a boundary is necessary to produce colours, and I said to myself aloud, as if instinctively, that Newton's theory is wrong.[1]

Due to circumstances, Goethe ended up looking through a prism rather than shining a ray of sunlight through it as Newton had done. What he saw was that colour only appeared on a boundary between light and dark. Although he initially concluded from this observation that Newton was wrong, he soon realised that it was not that simple.

In science a new theory normally needs to refute the accepted theory if it is to replace it. Yet soon after publishing his first study of colour, his *Contributions to Optics*, he realised that this does not disprove Newton's theory. This brought him to reflect on scientific methodology. Goethe realised that Newton's theory of colour is based not only on empirical experimental results, but also on what Goethe calls a *Vorstellungsart*, a way of seeing. The result of his methodological reflection is recorded in the short essay 'The Experiment As Mediator between Object and Subject', which he wrote in 1892 and is reprinted in this book. Here we see that, almost two decades before the publication of his *Farbenlehre*, Goethe had already abandoned his attempt to disprove Newton. Rather, he wanted to contrast Newton's way of seeing colour with his own way of seeing colour.

This distinction between two ways of seeing is important if we want to understand Goethe's approach to colour. But what does it mean to see something in a different way? Goethe's first look through the prism brought him to the realisation that a contrast is necessary for colour to arise. Goethe described this experience as an *aperçu,* which is French for 'glimpse' or 'insight'. It is the moment in which we see something new in something familiar, the moment in which a new aspect of something shows itself. However, Goethe uses this word in the third, historical part of the *Farbenlehre* to describe not only his experience of looking through a prism for the first time, but also Galileo Galilei's observation of the swinging church lamp:

> After science appeared forever fragmented by the Baconian scattering method, it was brought back together by Galileo. He led science back into the human being and by developing the laws of the pendulum and falling bodies from the swinging church chandelier he demonstrated already in his early youth that for the genius one instance is worth a thousand. In science everything comes down to what is called an *aperçu*, a becoming aware of what actually underlies appearances. And such an awareness is eternally fruitful.[2]

Goethe described the moment that led to Galileo's new way of seeing as an *aperçu,* and we find a wonderful description of this instance 'worth a thousand' in Thomas Kuhn's *The Structure of Scientific Revolutions*:

Since remote antiquity most people have seen one or another heavy body swinging back and forth on a string or chain until it finally comes to rest. To the Aristotelians, who believed that a heavy body is moved by its own nature from a higher position to a state of natural rest at a lower one, the swinging body was simply falling with difficulty. Constrained by the chain, it could achieve rest at its low point only after a tortuous motion and a considerable time. Galileo, on the other hand, looking at the swinging body, saw a pendulum, a body that almost succeeded in repeating the same motion over and over again ad infinitum. And having seen that much, Galileo observed other properties of the pendulum as well and constructed many of the most significant and original parts of his new dynamics around them. From the properties of the pendulum, for example, Galileo derived his only full and sound arguments for the independence of weight and rate of fall, as well as for the relationship between vertical height and terminal velocity of motions down inclined planes. All these natural phenomena he saw differently from the way they had been seen before.[3]

Galileo's new way of seeing brought about by his *aperçu* is what Kuhn calls a 'paradigm shift'. And Galileo's new scientific paradigm laid the foundation for modern science. Galileo's contribution to science, then, was not so much a refutation of the old Aristotelian view, but a new way of seeing.

Goethe's initial look through the prism was like the swinging church candles for Galileo. And just as Galileo's *aperçu* resulted in a new way of seeing, so did Goethe's. He remarked that the intention behind his *Contributions to Optics* was to bring about this *aperçu* in others so that they too could see colours as he did. His intention was not to refute Newton, but to inaugurate a new scientific paradigm. It is thus not a question of whether Newton or Goethe is right, but rather whether seeing colour as Goethe did brings us any closer to understanding colour in its diverse appearances.

The aim of this book is to guide you in seeing colours as Goethe did. When we travel through a foreign land to expand our horizons and broaden our minds, we cannot know in advance what it will be like or how it will change us. There is only one way to find out, and that is to begin the journey. It is the same with this journey through Goethe's world of colour. Only by walking in Goethe's footsteps and looking over his shoulder can we come to see colours as he did; by viewing the vistas that he saw and recorded in his *Farbenlehre* for posterity, and by viewing new vistas that have since been discovered using Goethe's way of seeing. Only by going on this journey through Goethe's world of colour will we know whether it brings us any closer to the essence of colour.

Figure 1.1: Cloud over a lake in front of the sun. The rays of light emanating from the sun are thousands of tiny images of the sun, one for each water droplet that reflects it.

1. Sight and Image

Goethe investigated the effects of light within the domain of colour phenomena, yet the role that light plays in human vision can also be explored in relation to brightness and darkness alone. In this chapter we will consider how visual phenomena in general arise, before turning to colour phenomena in the next.

Sight and light

We do not see light. Rather we see sources of light and the objects illuminated by them. Yet when the sun is shining through a cloud over a lake, do we not see rays of light streaming down? Let us consider this question.

When the cloud has passed over the lake we see the image of the sun reflected in its surface. When the wind blows across the water, the solar image splinters into a thousand dancing reflections. Previously, when we saw the sun's rays shining through the cloud, what we were seeing was not the light itself but thousands of little solar reflections, one for each droplet of water that reflected the light – the image of the sun – in our direction. So even in this example, we do not see light itself but rather miniscule images of its source.

Similarly, when we see objects – and even things like rainbows – what we see are actually thousands of reflections of the sun. However, due to the quality of the object's surface, those reflections are far more diffuse. Returning to our example of the lake, when a strong wind blows across it the waters become rough and all we can see is the surface. The reflections of the sun disappear into the appearance of the surface itself. The same thing happens with a matt surface: solar reflections are diffused across it.

Figure 1.2: Inside the cylinder it is dark (top). When you place your hand inside, it is illuminated by light from below (bottom). Without the presence of an object you cannot tell whether the light is on or off.

The invisibility of light

Light, then, is invisible, but it makes objects visible.

You can experiment with this outside on a dark night. If you shine a torch up at the sky, you won't see the light. But if you place your hand or an object in front of the torch, light will suddenly appear. If there is smoke or water droplets in the air, light will also appear. In each case, though, we are not seeing light itself, but reflections of the light source. We could say that light makes sight possible.

Presentation and representation

Light makes it possible for us to see. But what is it that we see? This question is not as straightforward as it might seem. Let us consider an example.

When we look at the sun, it is present in our visual experience of the world. The sun *presents* itself to us. But it is different when we see the shadow of a tree cast by the sun. The shadow does not only present itself to us as the sun did. It also represents something other than itself, in this case the tree. The shadow is not the tree, rather it is a *representation* of the tree.

There is a third phenomenon to consider. A small gap between leaves in the canopy of the tree casts a bright image of the sun in the shadow on the ground. Here the shadow no longer represents the tree, but the sun. During a solar eclipse we might notice that the shadows on the ground contain many bright crescent images. We had not noticed that the sun was being represented by the shadow too.

A small window can also cast a blurry image of the environment onto the wall opposite. Once we start looking out for this phenomenon we notice it more often than we might initially expect.

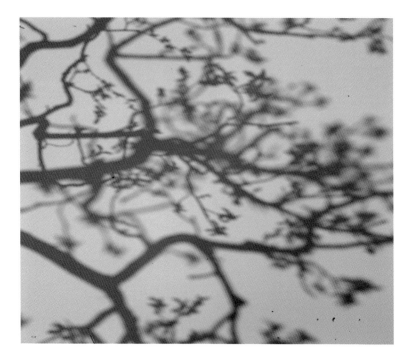

Figure 1.3: Shadows under a leaf canopy on a normal day (above) and during a partial solar eclipse (right). Even though the openings between the leaves have different shapes the crescent sun is clearly represented in the overlapping bright solar images.

Figure 1.4: Shadows cast by leaves. The leaves closer to the screen appear sharper, whereas those further away appear blurred.

Figure 1.5: Shadows cast on a wall by a crescent light source simulating your shadow during a partial solar eclipse.

Shapes of lights and shadows

You have probably noticed that as you approach a wall with the sun behind you, your shadow becomes sharper. If you walk backwards away from the wall, your shadow becomes blurred again. However, if you do this during a partial solar eclipse, you will notice that if you move your hand away from the wall, the shadow will appear increasingly less like your hand and more like many little crescent images.

The shadow is not only a representation of the shape of the object casting the shadow – in this case your hand – but also of the light source – in this case the crescent sun. If you move towards the shadow, the shadow will become sharper and more clearly represent you. If you move away, the shadow will become blurred and begin to represent the shape of the light source. Your shadow is therefore always a representation of you and the light source.

You can investigate the shadows cast by differently shaped light sources and shadow casters. For differently shaped light sources you can use a point light source and a straight light source, such as strip light or LED tube light. For shadow casters you can cut different shapes out of card. You will find that when the shadow caster is close to the wall, its shadow represents its shape. When it is close to the light source, the shadow represents the shape of the light source. When the shadow caster is in between, both its shape and the shape of the source are represented, creating an interplay of shapes. An annular eclipse, when the moon covers the centre of the sun and leaves only a visible outer ring, and the crescent moon are particularly interesting light source shapes to experiment with.

Shadow casters can also be inverted. Instead of using the shape you cut out of the card to cast a shadow, you can use the hole cut into the card. Observing the image produced by an aperture reveals the same effect as a shadow cast by an object, but with the role of light exchanged for that of darkness. You can see this clearly if you observe both situations next to each other: on the one side you will see a dark image in a light environment, on the other side a light image in a dark environment.

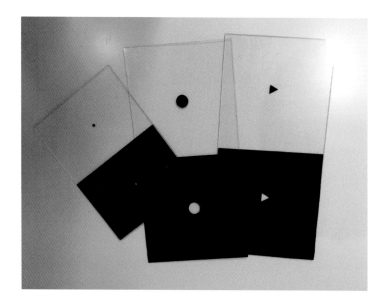

Figure 1.6: Inverse pairs of shadow casters made from card on glass (left). An annular light source (bottom left) and a crescent light source (bottom right) made from a circular light source.

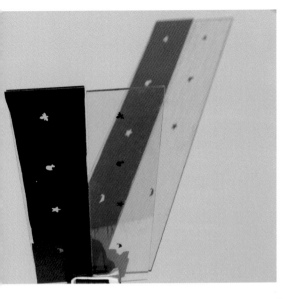

Figure 1.7: Shadows from inverse pairs of shadow casters. The further away the screen, the more blurred the shadows become.

Figure 1.8: The variation of the form of shadows can be investigated systematically using apertures and shadow casters mounted on a trolley. The shadows cast on the screen result from an annular light source.

The camera obscura

Camera obscura is the Latin term for a darkened room. You can make your own camera obscura by taping cardboard over the windows in a room. To create an image make a 5–10 mm (½ in) hole in the cardboard. Once your eyes have become accustomed to the darkness, which can take a few minutes, you will notice an image on the wall opposite. The image is an upside-down representation of the environment outside the room.

You can experiment with moving a screen closer and further away. You will find that the closer the screen is to the hole, the brighter and more blurred the image becomes. Conversely, the further away the screen, the darker and sharper the image becomes. You can also experiment with different sized holes. This will reveal the same relationship: the darker the image the sharper it becomes. However, as the image becomes darker, the colours disappear, just as at dusk the colours also disappear. By lightening the image by making the hole bigger or the screen closer you can experience the dawning of colour that accompanies the dawning of the day.

If you use tracing paper as a screen you can view the image from behind and thereby get very close to the aperture. By placing the tracing paper up against the hole and moving away you can experience the genesis of an image. You can also use the tracing paper to sketch the image. The camera obscura has now become a drawing aid. With your sketches you can investigate the laws of perspective that were discovered in this way in the Renaissance.

The first thing one invariably notices about the image created in the camera obscura is that it is upside down. But why is it upside down? If you now enlarge the opening in the cardboard so that it is big enough to see through, you will notice that although the image on the wall opposite has become blurred, there are still lighter and darker areas. If you look through the opening from one of the dark areas at the top of the image, you will see a dark area in the environment outside, such as the ground. The same applies for light areas, such as the sky at the bottom of the image. But if you want to see the sky through the opening, you will have to *lower* your head and if you want to see the ground you will have to *raise* your head, which is why the image is upside down. So what you see through the opening is represented on the screen at the position from which you see it.

In the camera obscura the environment acts as a light source and the upside down image on the wall is a representation of this light source. Just as with images of the light source in the previous experiment, a smaller aperture creates a sharper image. A lens has the same effect as a smaller aperture but allows more light and therefore a brighter image.

An opening in the overhead canopy of trees can create a natural camera obscura, which we can become aware of during a partial solar eclipse. Our eye is also a natural camera obscura; it creates a representation of the external world on the retina. If the eye did not form this image, we would not be able to see the world. So whenever we see something, there is a light source and a representation. The camera obscura in general and the photographic camera in particular can be considered as 'technical eyes'. We saw above that light makes sight possible. We can now add that to actually see, representation must occur.

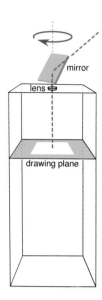

Figure 1.9: Diagram of a camera obscura drawing aid with a mirror and lens to represent the external environment on the drawing plane inside. The mirror can be rotated (red arrow) to select different views of the external environment.

Figure 1.10: Close-up of the drawing surface in the camera obscura drawing aid.

Figure 1.11: A camera obscura drawing aid representing the external environment on a drawing surface in the dark interior.

Figure 2.1: Look at the coloured image for 10-20 seconds, fixing your gaze on the black dot in the centre. Then look at the black and white image, again fixing your gaze on the black dot, and notice what happens to the image.

2. Colours Arising From the Eye

In the last chapter we saw that while light makes sight possible, representation of an image must occur in our eye in order to actually see. To understand how we see colour we therefore need to consider the function of the eye in more detail. In this chapter we will investigate colour phenomena related to the eye.

Successive contrast

You will have probably noticed that when you stare too long at the setting sun, a green image of the sun remains briefly in your field of vision when you look away. This trace of a previous observation is an afterimage. As the colour of this afterimage is different from that of the original object, there is a contrast between the two and, as this process happens sequentially, it is called 'successive contrast'.

Afterimages have been known since antiquity. Newton studied these appearances under the rubric of 'phantasms', and Goethe gave one of the first systematic descriptions of this phenomenon. We can say that the afterimage arises from the eye because the image detaches itself from the object we have just been observing and follows our gaze. This shows that our eye is not a passive organ. It reacts to what is seen in manifold ways.

Afterimages

Experiments with afterimages are the easiest experiments you can do. The colour arises from the eye so all you need is something to look at.

You can investigate afterimages systemically using coloured images on a white piece of paper. If you stare at a point inside a

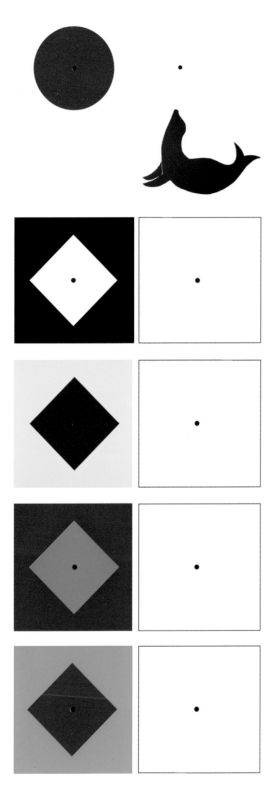

coloured image for a while without moving your gaze and, after a time, switch your gaze to a point on a white area, you will see an afterimage. The size and form of the afterimage remains the same as that of the original image, but the colour of the afterimage changes. The colour to which it changes is called the 'complementary colour'. Each colour has its own unique complementary colour. If you have a pattern containing two complementary colours, you can observe how the colours swap places in the afterimage. If you observe an object for a period of time, you will also notice a discolouration of the object as the afterimage is being formed.

As this is a physiological process the afterimage persists for only a short time. Usually, we do not notice afterimages because our gaze is constantly moving and they do not have time to form. Yet with careful observation you will notice this process occurring.

Figure 2.2: Look at one of the images for 10-20 seconds fixing your gaze on the black dot on the left. Then look at the dot on the right and see if the afterimage arises. Blinking sometimes helps.

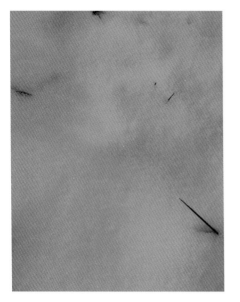

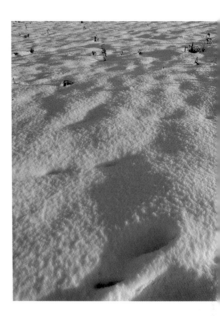

Simultaneous contrast

If you are outside looking at an object and a cloud passes overhead casting a shadow, the object you are looking at does not change colour. This phenomenon is so ubiquitous we normally do not notice it. However, if you are indoors at dusk and turn the light on, you might notice that the evening landscape outside has taken on a bluish hue. If you then go outside and look back into the room you were in, it will have taken on a yellowish hue. Also, in winter, if there is snow around and the sun is low in the sky, you might observe blue-violet or even greenish shadows on the snow.

These phenomena show that our eye responds to its environment by adapting to the colour of the illumination. This adaptation is a contrast phenomenon that happens much faster than afterimages, so it is called 'simultaneous contrast'. The simplest way to explore simultaneous contrast phenomena is by placing a square piece of grey card on different, uniformly coloured backgrounds. You will find that the environment of the card – the coloured background – has affected the appearance of the card itself; the eye has adapted to the environment of the object being observed.

Figure 2.3: Pale green (left), violet (middle) and blue (right) coloured shadows on snow at sunset.

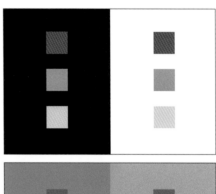

Figure 2.4: Squares in different environments. All the squares opposite each other have the identical colour or shade of grey – independent of their environment. But do they appear the same in different environments?

Coloured shadows

You might have noticed that a shadow cast by the setting sun can sometimes have a greenish tint. This can be particularly pronounced in a snow-covered landscape. In the first part of the *Farbenlehre*, the didactic part, Goethe describes the magical effect of the coloured shadows that appeared as he was descending the snow-laden Brocken, the highest peak of the Harz mountains, at sunset:

> At last the sun began to disappear and its rays, subdued by the strong haze, spread the most beautiful purple hue over my surroundings. At that point the color of the shadows was transformed into a green comparable in clarity to a sea green and in beauty to an emerald green. The effect grew ever more vivid; it was as if we found ourselves in a fairy world, for everything had clothed itself in these two lively colours so beautifully harmonious with one another. When the sun had set, the magnificent display finally faded into gray twilight and then into a clear moonlit night filled with stars.[4]

Goethe also describes how a candle and the full moon create the most beautiful orange and blue shadows.

Central to the phenomenon of coloured shadows is a second light source, one that is coloured. In the case of a snowy landscape, the sky illuminating the snow is the first light source and the coloured light source is the setting sun. Goethe did much of his experimentation on coloured shadows at night with two candles and a coloured piece of glass. Today we can use two projectors or two desk lamps as light sources. Place a coloured gel over one of the light sources and shine them both onto the wall. When a shadow caster is placed in front of the wall, both lights will cast a shadow, but although only one of the lights is coloured, both shadows are coloured. You will find that, as with afterimages, the colours of coloured shadows form complementary pairs.

Another situation in which you encounter coloured shadows is on the stage. If you stand on a stage that is lit by a green light and then a white light is added, your shadow cast by the green light takes on the hue of its complementary colour, namely magenta. However, if you walk towards your shadow so that it fills your field of vision, its magenta hue will gradually fade until it appears grey. This shows that your shadow only appears coloured in the context of an illuminated environment – it needs to be part of a visual whole.

We can understand coloured shadows by doing another simple experiment. Take a white object with you into coloured light. It will not turn the colour of the light but remain white. This shows that the eye adapts to the colour of the illumination by adding the complementary colour. If the illumination is green, the eye adds magenta. A shadow cast by the green light will appear dark, but if the shadow is illuminated by a white light so that it becomes a partial shadow, the eye will add magenta to that too. Instead of appearing grey, the shadow will appear magenta. Goethe was referring to this phenomenon when he remarked that the eye demands a totality.

Figure 2.5 (opposite): A white cup illuminated by a blue light (top) and additionally by a white light. The black shadow cast by the blue light becomes lighter, but instead of appearing grey it appears yellowish.

Figure 2.6 (left): The coloured shadows on a miniature stage cast by a green and a white light.

Figure 2.7: Circular pieces of card that have been coloured and made into spinning tops. The colours of the two tops that are spinning are beginning to mix.

Figure 2.8: A whirling table can be used to rotate contrast patterns and thereby discover a wide range of colour phenomena arising from the eye.

Moving contrast

Afterimages and coloured shadows are colours that arise from the eye when we observe stationary contrasts. But colours can also arise from the eye when we observe moving contrasts.

Cut a disc out of card, colour it with two colours, and stick a knitting needle through the middle or affix it to a board with a drawing pin. Then spin it. If you spin the disc fast enough, the two colours merge into one. Here, instead of a contrast in space giving rise to colours in the eye, a contrast in time does.

This type of contrast phenomena is sensitive to the frequency of the periodical stimulus, which, in the case of rotating discs, is determined by how quickly the disc rotates. Below around three revolutions per second the eye follows the moving pattern and no new colours arise. At higher frequencies the pattern begins to flicker and becomes blurred. All kinds of appearances arise which then vanish again before a stationary, single-coloured image emerges at even higher frequencies.

Benham's Disc

If you rotate a disc whose surface is divided into a black and a white section, you will see blurry, flickering patterns, which then merge into a homogeneous grey as you rotate it faster – but no colours. If you add small black arcs to the white section of the disc, the arcs will appear coloured when the white and black sections begin to merge. If you rotate the disc faster, you will see the arcs extend to form rings on a flickering grey background, each of a different colour. Reverse the direction of rotation and the order of the colours reverses.

Benham's disc is named after the British journalist and amateur scientist Charles Benham (1860–1929), who described it in the journal *Nature* in 1894. This phenomenon was first described by the French monk Benedict Prevost in 1826 and is still not fully understood today.

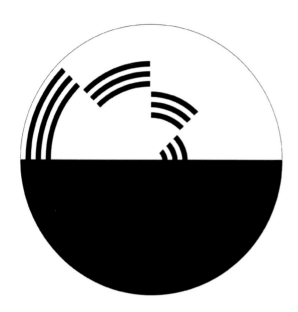

Figure 2.9 (top): Benham's disc. When the disc rotates rapidly, a spectrum of colours appears. When the direction of rotation is reversed, the order of the colours reverses.

Figure 2.10 (bottom): Motorised Benham's disc. The photograph shows what you might expect to see when the disc rotates rapidly. However, what you actually see is a spectrum of colours that cannot be photographed as they arise from the eye.

Colour mixing

If you spin a disc with segments of two different colours at a moderate speed, the colours mix and you no longer see the individual segments. If you mix blue and red you will get purple (figure 2.11), while yellow and magenta will give you orange (figure 2.12). You will notice that the colours produced have a lightness between that of the two original colours. You can experiment with other colour combinations; no doubt the process will remind you of mixing coloured pigments. If you mix complementary colours, such as magenta and green (figure 2.13), you will get grey – the colours have cancelled each other out.

Mixtures of more than two colours can also cancel each other out. For example, if you mix the colours of the rainbow (figure 2.14) you will also get grey. However, you will find that most combinations of colours will give another colour and not grey. More than two colours only mix to give grey when all the colours bar one mix to give the complementary of the remaining colour, so the idea of complementary colours cancelling also applies to mixing more than two colours.

Complementary colours cancelling each other out is a basic law of colour mixing. With coloured shadows we saw complementary colours arising together, now we see them disappearing together. These phenomena show us what it means for colours to be complementary.

Complementary colours

When white segments are added to a green disc, the two will mix to produce a light green when you rotate it. If white arcs are added that divide the green segments, you will see a white ring surrounded by light green (figure 2.15). When black is added to the arcs, spinning the disc does not produce grey, as you might expect, but a magenta hue (figure 2.16). As with coloured shadows, adding some darkness allows the complementary colour to appear.

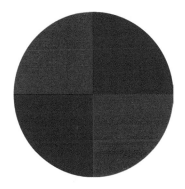

Figure 2.11: Rotating this blue and red disc produces purple.

Figure 2.12: Rotating this yellow and magenta disc produces orange.

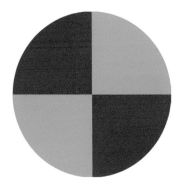

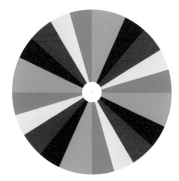

Figure 2.13: Rotating complementary pairs, such as these, produces grey.

Figure 2.14: Rotating the colours of the rainbow produces grey.

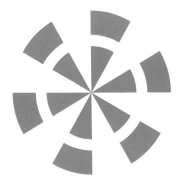

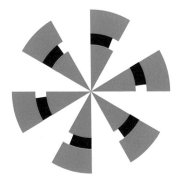

Figure 2.15: Rotating this green and white disc produces a light green colour, except for where the white arcs are. These produce a white ring in the light-green disc.

Figure 2.16: Rotating this green and white disc also produces a light green colour, except for where the black and white arcs are. These produce magenta, the complementary colour.

Figure 3.1: Atmospheric twilight colours.

3. Colours Arising From Colourless Media

The colour phenomena described in the previous chapter arise from our eye. As we will see in the next chapter, the colour we see when we look at a green leaf arises from the leaf's surface. But what about atmospheric colours? The reds of a sunset, the blues of the sky, the rainbow that appears if the sun shines when it rains. These colours arise neither from our eye nor from a surface, but from transparent, colourless media.

Unique to this group of colours is that their appearance depends upon our location and that of the light source. This means that where a rainbow appears depends upon where we are and where the sun is. If we move, the rainbow moves too. Likewise, the blue of the sky depends upon our location and the sun, although the colours change not when we move but rather when the sun moves. The different colours that arise depend upon the angle between the direction of the illumination and the direction in which we see the colour.

Turbid media

Atmospheric colours fascinated Goethe. Not just the blue of the sky and the yellows and reds of the setting sun, but also the blue cast of distant mountains. In order to understand these atmospheric colours Goethe investigated a related colour phenomenon that was later called the 'Tyndall effect' after the Irish physicist John Tyndall. You can observe this effect when a darkening substance is added to a transparent medium, such as smoke in the atmosphere. If you look at the sun through thick smoke it will look yellow, or even red. Even against the white background of a cloudy sky smoke will take on a yellowish hue. If you view the smoke against a dark

Figure 3.2: Smoke takes on a yellowish hue when seen against a light background, here the cloudy sky, and a faint bluish hue when seen against a dark background and illuminated from the side, in this case by the cloudy sky.

Figure 3.3: An opalescent glass bowl appears blue against the dark background and yellow against the white background. It has a yellow shadow.

background illuminated from the side by the sun or a cloudy sky, it will assume a bluish hue. These colours appear when a transparent medium becomes what we could call turbid, meaning hazy, cloudy.

Goethe was able to formulate a basic law of the illumination conditions under which specific colours arise, and of the changes of colour hues and saturation that depend on density and turbidity. Goethe explained the colours of the atmosphere in this way too. The yellows and reds of the setting sun and the blue of the sky arise in the turbid air of the atmosphere and depend on the angle of the sun. These atmospheric phenomena were first described mathematically in 1871 by the English physicist John William Strutt, Lord Rayleigh.

Opalescence

Solid turbid media, such as opalescent glass, give rise to colours in the same way that smoke does. For example, if you look at an opalescent glass object against a light background it appears orange, whereas if you look at it against a dark background when illuminated from the side, it appears blue. Goethe gave an empirical description of this phenomenon and how the colours change with the degree of turbidity. When the turbidity decreases, the shades of blue become more saturated and turn violet, whereas when the turbidity increases the shades of yellow become more saturated and redder. In this alternation of yellow and blue we can recognise the complementarity that we saw in the previous chapter.

Day sky and night sky

Watching the sky at dusk is a magical experience. The blue sky slowly darkens as the light of day gradually departs; the stars appear as one faint pinprick of light after another, forming constellations; and, finally, the Milky Way winds its familiar path across the heavens. As the blue of the sky disappears, the infinite universe is revealed to us in its unfathomable expanse.

You can recreate this phenomenon by filling a large glass container with water and adding a substance that makes the water very slightly hazy, or turbid (low-fat milk and soap work well). If you look through the water at something dark behind it, you can see the dark object, just like when you look through the atmosphere on a clear night and see the darkness of space punctuated with stars. If you illuminate the water from the side, it shimmers blue like the sky on a cloudless day. If you look at the wall, you will see that the illumination has become yellow where it passes through the water.

When the water is not illuminated the black background appears black through it. However, because the water is turbid, when illuminated from the side the black background is lightened to blue and the light shining through is darkened to yellow. Unlike opalescent glass, which can be so cloudy that it becomes translucent, the sky is only very slightly turbid and so transparent, allowing us to see the moon and the stars.

Figures 3.4 & 3.5 (opposite): An opalescent glass plate held up against a light background appears orange (left), whereas against a dark background it appears blue if illuminated from the side (right).

Figure 3.6: A column of water lit from below against a black background. Seen through the water, the black background appears a hazy blue like the blackness of space seen through the atmosphere during the day.

Figure 3.7: The water tank recreates the conditions of the atmosphere during sunset and sunrise.

Sunrise and sunset

As equally moving as the appearance of the night sky is the rising and setting of the sun. No two sunsets are the same and they vary from country to country. Near the equator the sun plummets into the sea, near the poles the summer sun struggles to dip below the horizon. Sunsets also change from season to season and where the sun crosses the horizon changes from day to day. The atmosphere, too, changes the quality of the sunset. Sometimes the sun barely changes to orange, sometimes it appears a deep, dark red. Not to mention the clouds that give each sunset its particular character. Observing the sunrise and sunset is something we never tire of.

You can recreate the sunset using a glass tank that is divided by a diagonal partition into two wedges. The wedge with the thickest end at the bottom is filled with slightly turbid water, the other wedge with clear water. If you look at a light source through the top of the tank, it will appear pale yellow. If you bend down, the light source appears to sink lower in the tank turning dark yellow, then orange and possibly red. You have recreated what happens when sunlight passes through ever thicker layers of the atmosphere as the sun sets, except in this case it is not the sun that is moving, but you. When you stand up straight again, you recreate the effect of the rising sun.

If you walk round to the side of the tank, the turbid water will now look dark blue – almost indigo – at the top of the tank, becoming a less saturated blue further down until it reaches a pale greenish-blue at the bottom. From this angle, you are seeing what happens when you look up at the blue sky with the blackness of space behind it.

Because the turbid water is wedge shaped, its thickness increases from top to bottom, recreating the increasing thickness of the atmosphere that we look through when we gaze from the zenith down to the horizon. This arrangement of two wedges in the tank allows for a turbid wedge that does not produce the coloured fringes that a single prism would; here the prismatic effects of the two wedges cancel each other exactly.

This simple experiment shows not only two continuous

sequences of colour, but also their mutual relationship. Each colour in a sequence is complementary to the corresponding colour in the other sequence: yellow appears at the top of the tank, as does dark blue or even violet, and red appears at the bottom, as does light blue or even cyan. Moreover, the sequence from yellow to red is due to the turbid medium increasingly darkening the light and the sequence from blue to cyan is due to the obliquely illuminated turbid medium increasingly lightening the darkness. So we could say that the first sequence arises by progressively darkening the light and the second by progressively lightening the darkness. Inverting this relationship between light and darkness transforms a colour in one sequence into its complementary colour in the complementary sequence.

The colours of the blue sky and the setting sun thus show the lawfulness of their arising in the relation between light and darkness. Goethe felt that these atmospheric colours display most clearly the lawful arising of colours in general, and therefore referred to the blue sky and setting sun as an 'archetypal phenomenon' in his *Farbenlehre*.

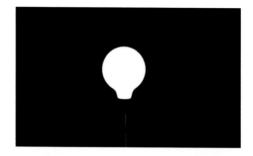

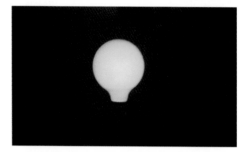

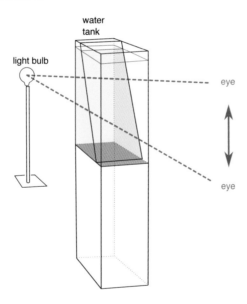

Figure 3.8: Diagram of the lightbulb seen through the water tank to recreate the conditions of the sunset and sunrise. Moving the eye up and down (red arrow) changes the thickness of the turbid medium (blue) through which the lightbulb is seen.

Figure 3.9: Simulated sunset and sunrise. A direct view of the lightbulb (top), which represents the sun. The lightbulb observed through the top of the tank (top middle) is yellow and turns orange as it is viewed through a thicker turbid layer (bottom middle and bottom). This demonstrates how the yellow and red colours appear with increasing atmospheric turbidity.

Figure 3.10. Prismatic colours arising from water droplets reflecting and refracting the sunlight.

Prismatic colours

Colours arising from the eye depend not so much on the spatial configuration of the surroundings, but on the eye itself. This is most apparent for afterimages: wherever you look, the afterimage follows your gaze, it is completely independent of your surroundings. In colours arising from colourless transparent media, the spatial configuration plays a leading role. The colour of the setting sun, for example, depends on the angle between you, the sun and the horizon.

When you look into a swimming pool at an oblique angle, the tiles appear the same colour as they do above water, but coloured fringes appear at dark edges. The same happens if you look through a prism. A white wall will appear unchanged but coloured fringes will arise where there are contrasts between light and dark areas. Prismatic colours arise when you view such a contrast obliquely through a dense, transparent medium such as water or glass.

You can also project colours arising from a prism onto a screen using a light source to create a contrast. You will find that prismatic colours depend even more on geometrical configuration and that distance plays an important role too: changing the angle of the prism by a few degrees or moving the distance of a screen by a small amount can greatly change the colour that appears at a given position. Newton projected the image of the sun through a prism onto a screen in his darkened chamber to develop his theory of light, which we will consider in Chapter 5.

Elevation and refraction

If you have ever been snorkelling in the sea, you will have noticed that the seabed looks closer than it is. Sometimes it seems like you can touch it, but as you reach down, your hand does not even come near.

You can investigate this observation with an aquarium half-filled with water and placed on a chequered pattern. If you look into the aquarium directly from above, the pattern at the base appears elevated. Looking from an oblique angle by stepping backwards elevates the base even more and it becomes deformed. In addition, coloured fringes not seen when looking directly from above appear at the boundaries of the white and black pattern.

Seen from above, a rod placed diagonally into the aquarium appears to bend upwards. Optically, the rod is elevated like the base. Seen from the side, the rod appears straight. By observing different objects from different angles you can observe that when the angle of view moves away from the perpendicular, the object appears increasingly elevated, it is displaced towards the water surface along the perpendicular. As the displacement towards the surface increases, the size of the coloured fringes also increases.

A denser medium such as glass can be used instead of water, which increases the displacement. Also, instead of looking through a boundary between two transparent media, an image can be projected through the boundary onto a screen. The displacement of the position of the image on the screen is called 'refraction'.

Figure 3.12 (bottom left): An aquarium half-filled with water is mounted on a black and white pattern. The chequered base appears elevated when seen through the surface. The rod, which appears straight when seen through the side, appears to bend upwards when seen through the surface.

Figure 3.13 (bottom right): Close-up of the rod appearing bent when seen through the surface. Faint coloured fringes can also be seen at the boundaries of the chequered pattern.

Figure 3.11: The water in a lake appears shallower than it is, making objects in the water appear shorter than they are.

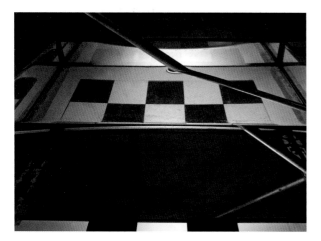

Figure 3.14 (top): Black and white stripes behind a convex lens. The stripes look bigger through the lens - magnification occurs. Also, coloured fringes appear on the boundaries away from the centre of the lens.

Figure 3.15 (bottom): A mounted concave lens with a chequered black and white board underneath. The plate looks further away when seen through the lens - minification occurs.

Convex and concave lenses

If you are out walking in nature after it has been raining, you might notice how everything sparkles and glistens due to droplets of water nestled in the grass and leaves. If you observe a waterdrop close up, you will notice that it magnifies the structure in which it is nested; it has become a magnifying lens.

Lenses are another example of refraction. Observe a chequered pattern from roughly 20 cm (8 in) through a large convex and then concave lens. As you move the lens away from the pattern, a magnification (convex lens), or minification (concave lens) of the pattern occurs. As you move the lens closer the effect diminishes. Coloured fringes will appear at the black and white edges if you look through the lens at an angle.

We have seen that water displaces the object we look at. Here we see that glass does that too. The convex lens is thickest at the centre and tapers outwards. It magnifies the image by displacing each part away from the centre. Conversely, the concave lens is thickest at the circumference and tapers inwards. It minifies the image by displacing each part towards the centre. Binoculars, microscopes and projectors all utilise this effect.

This relationship between magnification and distance with convex lenses only holds when you view nearby objects with your eye close to the lens. If you view an object through a magnifying glass and then move away from the magnifying glass, the object will get bigger until you reach a point where the image turns upside down and then gets smaller as you move further away. You can observe small upside-down images in raindrops.

The prism

When you look through a lens, parts of the image are displaced either towards the centre or away from the centre. This magnifies or minifies the image, but it does not displace it *as a whole*. A prism is an elongated section of a lens. When you look through it, you see that the image is neither magnified nor minified but displaced as a whole in the direction of the thick end. As with the lens you can also observe a distortion of the image and the appearance of coloured fringes where there is a contrast between light and dark.

Begin experimenting with a prism by simply looking at the world through it. As we saw in the Introduction, this is where Goethe started too. It doesn't matter what kind of prism you use, but with a large water prism the displacement of the image and the coloured fringes become especially clear. The first thing you will notice is that you can only see prismatic colours when an image is displaced. However, if you look at the displaced image of a uniformly coloured surface, like a green meadow or grey sky, you will not see any colours. For prismatic colours to arise, not only is a displacement necessary, but also a contrast.

Figure 3.17: A water prism can be made by gluing together four pieces of glass or Perspex: two pieces for the sides and two for the wedge. The image seen through the tank is displaced downwards as a whole and coloured fringes appear where light and dark areas contrast.

Figure 3.18: When you look at nature through a prism, you will see coloured fringes appear where light and dark contrast.

Figure 3.16: Raindrops are also a convex lenses. When the object is very close raindrops magnify the object (far left). When the object is not very close the image turns upside down and is minified (left).

You can investigate prismatic colours further by looking through a prism at black and white patterns.

The first observation you can make is that colours only arise at a boundary between black and white that is perpendicular to the direction of displacement. With this simple pattern, the direction of displacement determines which colours appear. Yellow and red arise when a black surface is displaced in the direction of a white surface (figure 3.19), whereas cyan and blue arise when a white surface is displaced in the direction of a black surface (figure 3.21). You can change the direction of displacement by turning the image upside down, which inverts white and black. We encounter once more the two colour sequences of the atmospheric colours, although this time the intermediary colours occupy only a very thin band. Again, inverting light and darkness transforms one sequence into the other.

You can place these two situations side by side (figure 3.22). The original opposition of black and white now includes yellow opposite blue, and cyan opposite red. In this situation, colours arise in pairs. These pairs are the familiar complementary colours, and we can see that complementary colours are also complementary with respect to light and darkness: yellow and cyan are light colours and appear next to white, whereas red and blue are dark colours and appear next to black.

Figure 3.19: Look at the pattern on the left through a prism with the apex facing downwards. Red and yellow arise when a black surface is displaced in the direction of a white surface.

Figure 3.20 (left): Tetrahedral water prism showing different views of the room. These are produced by images being displaced in different directions by different surfaces of the prism.

Figure 3.21 (below): Cyan and blue arise when a white surface is displaced in the direction of a black surface.

Figure 3.22: The configurations of figures 3.19 and 3.21 are now next to each other. Yellow appears opposite blue, and red opposite cyan. In this configuration, colours arise in their complementary pairs.

So far we have not seen all the colours that appear when looking through the prism at the world. Specifically, green and magenta are missing. These two colours arise by bringing the two edges together. This you can do by looking at a light strip on a dark background, first from close up and then moving away with the prism (figure 3.23). As you move the prism away, the cyan and yellow come together and overlap. Where they overlap green appears. Here we have two light colours mixing to give a darker colour. If you keep moving away, the cyan and yellow overlap completely and disappear into the green. Moving away thus darkens the spectrum until only red, green and blue remain.

You can do the same again, but with light and dark inverted (figure 3.24). Here, where the blue and red overlap, magenta appears. This time, however, we have two dark colours mixing to give a light colour. If you keep moving away, the blue and red overlap completely and disappear into the magenta. Under these conditions moving away lightens the spectrum until only yellow, magenta and cyan remain.

Figure 3.23: When you look at this pattern with the prism up close and then move away, the cyan and yellow will come together and overlap. Where they overlap green appears. If you keep moving away, the cyan and yellow will overlap completely and disappear into the green.

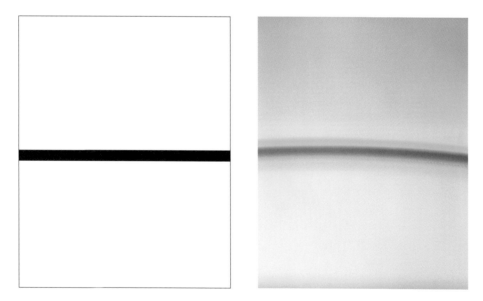

Figure 3.24: This pattern is the inverse of figure 3.23. Now, where the blue and red overlap, magenta appears. If you keep moving away, the blue and red will overlap completely and disappear into the magenta.

Figure 3.25: The two previous configurations are now next to each other. This produces two sequences of complementary colour pairs.

Like with the afterimages in the last chapter, the prismatic colours arise in complementary pairs. If we place the two situations side by side (figure 3.25), we now have the following two sequences opposite each other:

White	Black
Cyan	Red
Blue	Yellow
Magenta	Green
Red	Cyan
Yellow	Blue
White	Black

When we place these two sequences side by side, we see that they are all the complementary pairs that we saw in the last chapter. So not only do colours arising from the eye arise in complementary pairs, but also from colourless media.

Neither of these two colour sequences contain all six prismatic colours. Each sequence is missing one colour that the other one has. So if we want to represent all the prismatic colours, we need to bring the two sequences together into a circle:

Magenta

Red   Blue

Yellow   Cyan

Green

This colour circle, then, represents the relationships between the prismatic colours: complementary colour pairs are diagonally opposite, colours that are adjacent in the edge spectra are likewise adjacent in the colour circle, and the two colours produced by combining the two edge spectra are opposite each other at the top and bottom of the circle between the two colours that overlap to produce them (red and blue for magenta, yellow and cyan for green). We also find the colour sequence of the sunrise represented on the left (upside down) and the colour sequence of the blue sky represented on the right.

Figure 3.26: Close-up of the window frame seen through the tetrahedral water prism in figure 3.20. The coloured fringes are especially clear.

Figure 3.27: The double rainbow.

The rainbow

The rainbow is one of nature's most spectacular colour phenomena. It appears when the sun is behind you and shining into a curtain of rain in front of you. Its appearance is an interplay between the sunlight and countless falling raindrops. Where the rainbow appears depends on where the sun is in the sky. It also depends on where you are standing. In fact, the rainbow will form a circle around the shadow of your head. It thus appears at a different location for each observer. You can experiment with the rainbow using a hosepipe to create water droplets on a sunny day. You will find that again the rainbow appears when the sun is behind you and the water droplets in front. You can even look first through one eye and then the other to see a slight shift in the position of the rainbow.

To understand how the rainbow arises, you can experiment with a single artificial raindrop: a glass sphere (a wine glass filled with water will also work). If you illuminate the sphere horizontally using a circular light source in a darkened room, you will recreate the sun when it is low in the sky. The light is reflected from the internal surface of the glass sphere onto a white wall. Colours arise at the boundary between light and dark forming an arc, just as they do in nature but slightly smaller because glass is denser than water and so displaces the image more. With this setup you can step into the rainbow and look back into the sphere. You can see that the rainbow is made up of thousands of overlapping coloured images of the light source. This is what happens when you are looking at a rainbow in the sky: you are seeing a multitude of tiny images of the sun, each created by a raindrop.

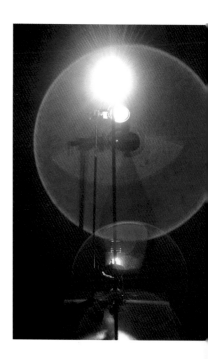

Polarisation colours

The turbidity of the atmosphere produces the most spectacular atmospheric colours. As we saw, these colours depend greatly on spatial configuration, such as where the observer is standing in relation to what they are observing. There is another atmospheric phenomenon that you can observe particularly well at sunset. If you are watching the sunset from a boat on a still lake or from the shore, pay attention to the water nearest to you. You will notice that, rather than reflecting the blue evening sky, the water appears eerily dark and deep – the image of the overhead sky has disappeared. During the day, the water simply reflects the blue sky, but during sunrise and sunset the reflection of the blue overhead sky disappears, whereas the reflection of the rest of the sky, and any overhead clouds, remains unchanged. In order to understand why the reflection of the overhead sky disappears, we need to consider a related phenomenon that can be observed with Iceland spar, a transparent calcite crystal.

Figure 3.28 (top left): Shining a light on a glass sphere creates a rainbow.

Figure 3.29 (top middle): A glass sphere is illuminated horizontally by a mirror that reflects light from a light source below. A black board is added for viewing the rainbow behind without being dazzled by light from the glass sphere.

Figure 3.30 (top right): The light from the mirror is reflected from the internal surface of the sphere onto the white wall. A rainbow appears at the boundary between light and darkness.

If you observe a pattern through Iceland spar, two images of the pattern will appear. They often look elevated by different amounts or have different colours. Each of the two images displays the familiar characteristics of refraction seen in the previous section. This double displacement of the image arises due to the double refraction that occurs in calcite crystal, which makes it a 'birefringent medium'.

Transparent media are birefringent when their internal structure is what is known as anisotropic. This means they exhibit different optical properties when viewed from different directions. Conversely, water and glass have the same optical

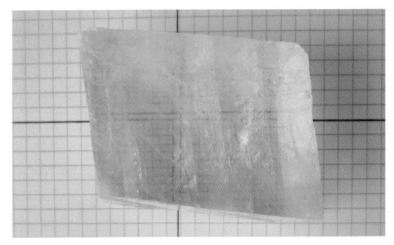

Figure 3.31 (above): When the sun is low in the sky the reflection in the water of the blue sky overhead disappears. The turbid atmosphere perpendicular to the sun is the first polariser and the reflecting surface of the lake the second.

Figure 3.32 (top right): A pattern viewed through Iceland spar, a transparent calcite crystal, appears as two images elevated by different amounts. This is a case of double refraction in a birefringent medium.

Figure 3.33 (bottom right): A black rectangle viewed through Iceland spar creates two images, one green and one magenta. They appear black where they overlap.

properties when viewed from different directions. They are therefore isotropic media: transparent media that create a single displaced image by refraction.

So far, we have observed two seemingly unrelated phenomena: a disappearing reflection of the blue overhead evening sky in a transparent medium (water), and two images created by double refraction in an anisotropic transparent medium (Iceland spar). You can see what these two phenomena have in common if you then look at what happens to the two images when the Iceland spar is reflected by a transparent medium: one of the images becomes faint and even disappears. In both phenomena, then, a reflected image disappears. In the case of the blue overhead evening sky the disappearing image leaves the water looking deep and dark, whereas in the case of the double image the disappearing image leaves the other image more or less unaffected.

Because one of the images seen through Iceland spar disappears when reflected, but not the other, there must be a difference between the two. Moreover, which image disappears depends on the orientation of the Iceland spar to the reflecting surface and you can observe the reflected images disappearing and reappearing by rotating the Iceland spar. You will find that one reflected image disappears and the other reappears with each 90 degree rotation – they are opposed to each other with respect to their appearance and disappearance in the reflection. We could thus say that the two images are polar to each other and that they are polarised by the transparent calcite crystal. In birefringent media, then, refraction and polarisation occur at the same time.

The blue overhead evening sky and one of the images produced by Iceland spar disappear when reflected. So, analogous to the polarised images of Iceland spar, we can say that light from the overhead evening sky is also polarised. But why does a polarised image reflected by a transparent medium disappear?

In the previous example, the reflection of Iceland spar was observed in a transparent medium. But you can exchange the position of the Iceland spar and the reflecting surface by taking

Figure 3.34: A black rectangle viewed through a transparent calcite crystal (top) creates two images. When reflected by a transparent medium (bottom) one of the images disappears.

the crystal and looking through it at a reflection in a transparent medium, such as water or glass. By rotating the crystal, you will notice that reflections disappear and then reappear with the changing orientation of the crystal.

This observation, in which the position of Iceland spar and the reflecting surface have been swapped, shows that the reflecting transparent medium is also a polariser, and that the polarisation becomes apparent when the two polarisers are arranged in tandem. In the previous observation, Iceland spar was the first polariser and the reflecting surface the second; now the reflecting surface is the first polariser and Iceland spar the second.

We can use these observations with Iceland spar to understand our first observation. The overhead evening sky is a transparent medium that polarises light from the sun. We see that the light is polarised when it is polarised a second time by the second transparent medium (the water) and thereby darkened. This phenomenon also shows the importance of the geometrical orientation for two polarisers to darken the light: the sun must be at the horizon for the overhead sky to be polarised in such a direction that the reflection is darkened by the horizontal water.

We saw that Iceland spar creates two polarised images by birefringence, yet the reflecting water only seems to create one image, the reflection. So if the reflection of, say, the sky is indeed polarised, where is its polar image? In order to see the second image we must jump into the water and look up. We can see the sky from under the water too – that is the second image. One image is reflected by the water's surface, the other is transmitted into the water.

It is important to note that in the different examples we have considered, the transparent calcite crystal, the water and the overhead evening sky all polarise light by different means, namely by birefringence, partial reflection and turbidity respectively. Yet the general phenomenon of polarisation is the same in each case: two polarising transparent media, in a certain orientation, darken the light.

Polarisation colours with a reflecting surface

Two polarising, transparent media, correctly aligned, darken the light. However, when a birefringent medium is placed between the two polarisers you can see it – it appears light against the dark surroundings. Moreover, it will also appear coloured. This is because birefringent media create multiple images, as we have seen, and these images are now no longer fully aligned with the second polariser and therefore not completely darkened.

As we saw, the evening sky and a water surface both polarise light, so to produce polarisation colours all you need to do is place a birefringent medium in between. If you don't have a body of water nearby, a thin layer of water in a black tray will work fine. Place different transparent objects above the water and see if their reflection is coloured.

Generally, plastic objects will appear coloured and glass ones will not. Glass, as we have seen, is normally isotropic. But the way plastic objects are produced causes internal tensions, and tensions in transparent media makes them anisotropic. Particularly colourful is Sellotape layered on a sheet of glass. Placing layers on top of each other changes the colour, so having layers of different thicknesses causes different colours to arise.

In order to produce polarisation colours in this way, you will need to experiment with which part of the sky is reflected in the water and the time of day. You will discover that light from the area of the sky that is at right angles to the sun works best, which is the darkest part of the sky. Also, you will find that mornings and evenings work best. This is why the darkening of the reflected overhead sky can be observed most clearly at sunrise and sunset.

Figure 3.35: Colours appear in the reflection of a transparent plastic jug.

Figure 3.36: Layers of Sellotape on a glass sheet appear coloured in the reflection.

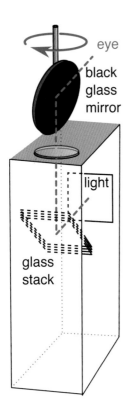

Polarisation colours with two reflecting surfaces

We have seen that polarisation colours arise when birefringent media are placed between two polarisers. It is also possible to use glass as polarisers instead of the sky and water, because reflections from any transparent media are polarised. In this experiment, the sky as a first polariser is replaced by a stack of transparent glass plates. The water as a second polariser is replaced with a black glass plate (any reflecting medium can be used, but metal surfaces, including mirrors, will not work). The inclined transparent glass plates reflect a light source upwards, which is then reflected by the inclined black glass plate such that you can see a reflection of the light source in the black glass plate.

The black glass plate is rotatable so that you can investigate the relation between the two polarisers. When the black glass plate is parallel to the transparent glass plates, the light source appears bright. If you rotate the black glass plate until it is perpendicular to the transparent glass plates, the light source will become darker until it is fully dark. If you place a birefringent medium between them in this orientation, you will see that the object's reflection is coloured. If you rotate the black glass plate so that the background changes from dark to light, the colours in the birefringent medium change to their complementary colours. The colours that arise here are related to light and darkness in the same way as the colours that arise in a turbid medium and a prism: inverting light and darkness changes the colours into their complements.

Figure 3.37 (top): Diagram showing the arrangement of a stack of transparent glass plates and a rotatable black glass plate that both reflect a light source. Birefringent media are placed in between.

Figure 3.38 (bottom): Tempered glass, a birefringent medium, illuminated from below by a light source reflected by a stack of transparent glass plates. Polarisation colours appear in the reflection of the tempered glass in the inclined black glass plate.

Transparent media are birefringent when their structure is anisotropic and therefore produce polarisation colours when placed between two polarisers. Crystals have naturally occurring anisotropic structures, but it is also possible to induce such structures. Ordinary or annealed glass, such as a windowpane that has not been heat-strengthened, is isotropic and so no colours arise. If glass is heated and then cooled unevenly, such as tempered glass, internal tensions form in the glass and it becomes anisotropic. One can also place a piece of glass under tension by exerting an external pressure on the glass with a clamp. Brilliant colours arise if you then view it between two polarisers. As internal tensions in a transparent medium cause birefringence, the polarisation colours that arise are used to analyse internal forces and structures.

Polarising gels

When you look through polarising sunglasses and tilt your head, the reflections will change – some become lighter, some darker. This is because the glass has been coated with a polarising film. You can explore the phenomenon of polarisation using polarising films, such as polarising gels, instead of Iceland spar or reflecting surfaces.

Begin by experimenting with a single polarising gel. When you look through it, everything seems fairly normal, but as you rotate it, reflections in water or glass surfaces appear and then disappear and the sky changes from dark blue to pale cyan and back again. This will remind you of looking through Iceland spar and rotating it.

Now experiment with looking through two polarising gels. When they are aligned so that you can see through them, everything looks the same as when looking through the single polariser. This corresponds to the parallel alignment of two reflecting surfaces. If you rotate one of the gels by 90 degrees, the view turns more or less completely black, just as it does when two reflecting surfaces are perpendicular.

This helps us understand what happens when we look through a single polarising gel. As we have seen, transparent and turbid

Figure 3.39: Reflections on a water surface seen though a polarising gel (top). When the gel is rotated by 90 degrees, reflections disappear (bottom).

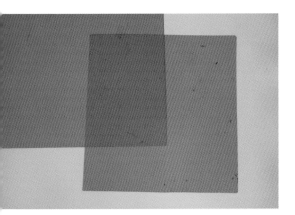

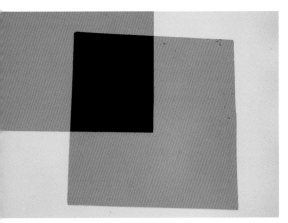

media polarise light when they reflect it, so when we look through a polarising gel, such as polarising sunglasses, it acts as the second polariser, darkening reflections and the sky, which are polarised, when it is rotated. Polarising gels are thus like Iceland spar, and although they also polarise by birefringence, you don't get two images when looking through because one of the images is absorbed.

You can create a polarising sundial by cutting a polarising gel into twelve segments of a circle with the direction of polarisation the same for each segment. When the individual segments are placed on top of each other in the same alignment, minimal darkening occurs and you can see through them easily. If you arrange them in a disc on a window, different segments darken the sky by different amounts. This will change not by rotating the disc, but throughout the day as the position of the sun changes. As the darkened segments of the disc follow the sun throughout the day, the disc acts as a sundial.

Figure 3.41: Segments of polarising gels fixed to a window. As the sun changes position throughout the day, different segments are darkened, creating a sundial.

Figure 3.40: A light background seen through two polarising gels when they are parallel (top), and when one is rotated first by 45 degrees (middle), and then by 90 degrees (bottom).

Polarisation colours with two polarising gels

If you put on polarising sunglasses and view a white area on a computer flatscreen through a thin plastic film, such as clingfilm, you won't see any change. But if you stretch the film colours will arise, because stretching causes internal tensions and makes it anisotropic. If you tilt your head left and right, the colours will change. The computer flatscreen contains a polariser, so you are viewing an anisotropic medium between two polarisers.

You can explore polarisation colours with one rotatable polarising gel mounted above another and illuminated from below. If you rotate the top gel as you look through it, the light source will turn dark and then light again. If you then place birefringent media, such as Sellotape, cellophane, mica, plastic, tempered glass or a film of liquid crystals, between the two gels, they will appear coloured. The colours change when you rotate the gel: when the background changes from light to dark, the colours turn into their complementary colours. This is similar to interchanging light and darkness in the prism experiments. However, an important difference is that prismatic colours only arise at contrast edges, whereas polarisation colours arise because the medium is anisotropic and therefore birefringent.

Figure 3.42: A thin plastic film stretched in front of a white computer flatscreen seen through polarised glasses.

Figure 3.43 (right): A film of liquid crystals is placed on a polarising gel illuminated from below. If you view it through polarised glasses and apply pressure to the gel, you will see the colours change. You can also heat the liquid crystals using a hair dryer. At a certain temperature they turn fully transparent because they lose their internal structure and become isotropic. As they cool, their anisotropic structure is restored - they become birefringent again - and colours reappear.

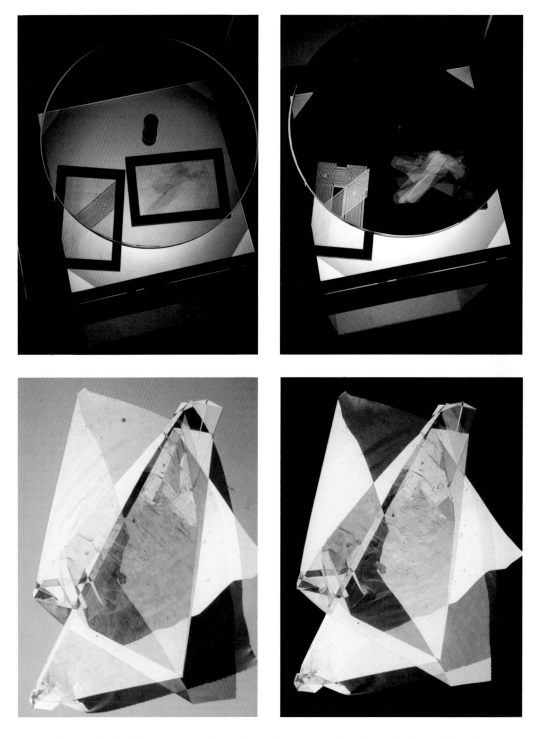

Figures 3.44 & 3.45: Sellotape (top) and cellophane (bottom) illuminated from below and viewed between two polarising gels. The gels are parallel (left) and perpendicular (right).

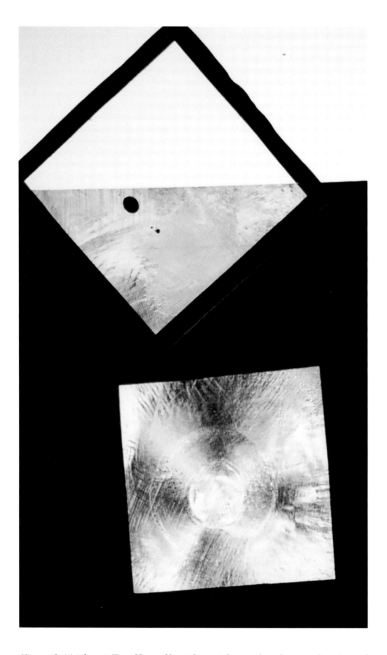

Figure 3.46 (above): Two films of liquid crystals are placed on a polarising gel illuminated from below. The darkened area is due to its being viewed through a second, perpendicular polariser. Only here do polarisation colours arise.

Figure 3.47 (right): Mica illuminated from below and viewed between two polarising gels. When the top gel is rotated from being parallel (top) to being perpendicular (bottom) with the bottom gel, the colours change to their complements.

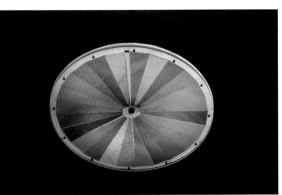

Figure 3.48: A rotating disc made of segments of polarising film covered by a birefringent film and lit from behind making it appear shining white (top). The reflection of the disc on the water's surface is coloured and the segments change colour to maintain the orientation of vertical yellow and horizontal blue segments. Close-up of the reflection of the rotating disc (bottom).

Polarisation in the eye

We saw that polarisation phenomena only arise when there are two polarisers. In the phenomena considered so far, both of these polarisers are outside us; for example, the evening sky overhead darkened by the reflecting water. However, if you look at the dark area of the sky and slowly tilt your head from left to right, you might be able to notice a small, fuzzy yellow bar appear in the sky as long as you keep your head moving. The faint yellow bar looks like it is painted with a brush, which is why it is called 'Haidinger's brush', after the Austrian mineralogist Wilhelm Karl von Haidinger (1795–1871), who discovered the effect in 1844.

In addition to the yellow 'brush', even fainter bluish and purplish 'brushes' can appear above and below it. You can also see it by tilting your head left and right while looking at the white area of an LCD computer screen, which, as we have seen, is also a source of polarised light. The orientation of the brush depends on the relation of the area of sky to the sun, or the orientation of the polarising film in the screen.

Polarisation colours appear when a birefringent medium is viewed between two polarisers. Yet with Haidinger's brush, only a single polarising medium is required and no birefringent medium. This is because the second polariser and the birefringent medium are in the eye itself. Haidinger's brush is a kind of afterimage created by the eye in response to viewing light through a polarising medium.

It is possible to represent externally what is happening in the eye. This experiment, designed by German physicist Johannes Grebe-Ellis, has two polarisers: a rotating disc, lit from behind and made up of segments of polarising gels that represent the polarising properties of the macula at the centre of the retina, and a surface of water. The segments of polarising gels are covered with transparent plastic, which is birefringent due to stretching in the manufacturing process. This layer represents the birefringent properties of the cornea. The polarisation of the water surface represents the polarisation of the sky. Lastly, the rotation of the disc represents the movement of the head.

When you observe the reflection of the rotating disc in the water, you can see that the direction of rotation is reversed. It also has an intense, cross-shaped pattern of complementary colours with vertical yellow and horizontal blue-violet segments whose orientation remains unchanged as the disc rotates.

Diffraction colours

Colours arise from colourless transparent media in diverse ways, yet we can find examples of nearly every kind in the atmosphere. The blue of the sky is an example of colours arising from turbid media and the rainbow is an example of colours arising from transparent media. Both are also polarised, which shows how closely related these two kinds of phenomena are.

There is another class of phenomena that finds expression in the atmosphere – diffraction colours. You might have seen a corona appear around the sun or moon when viewed through water droplets in a cloud. It is also possible to see a colourful corona around the sun in spring when pollen drifts through the air. If you observe such a corona, you will notice that the circular shape does not correspond to the shape of the sun, but to the orientation of the pollen grains: if they are oblong and orient themselves in the same direction by slowly falling when the wind is still, the corona will become elliptical. The corona becomes wider due to the narrow width of the pollen grains and shorter due to their height.

If you view a light source through fine patterns such as a feather or a spider's web, diffraction colours also arise. You can create diffraction colours by scratching parallel lines on the surface of a glass plate to create a fine structure. If the structure is fine enough, periodic colours – colours that repeat at regular intervals – will arise.

Figure 3.49 (top): A corona around a light source viewed through evenly distributed lycopodium powder (spores of a fern-like plant). The periodic nature of diffraction colours is clearly visible.

Figure 3.50 (bottom): A light source viewed through a feather creates diffraction colours.

Figure 3.51: The thin silk strands of a spider's web illuminated by the sun create diffraction colours.

Colours seen through fine patterns

Place a candle about 3 m (10 ft) away in a dark room. If you bring your hand close to your eye and make a small gap between your forefinger and thumb you will see colours when you look at the candle through the gap. You can experiment with colours arising through fine patterns using other light sources, such as a small halogen light and different coloured LEDs. You can view these light sources through finely structured materials such as scratched plates, lycopodium powder (made from the spores of a fern-like plant), fabrics, a bird feather or a grating with narrow parallel lines.

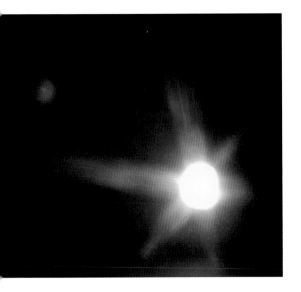

Figure 3.52: A circular light source viewed through a glass plate with parallel lines scratched on it in three different directions.

Figure 3.53: A selection of fine patterns. If you view a light source through fine patterns, diffraction colours appear.

If you view these patterns from close up and focus on the light source, you will see coloured patterns around the light. Experimentation shows that the smaller the structure, the larger the colour appearance, and the more regular the structures are, the more regular and saturated the colours become.

These colours are examples of diffraction colours, which occur periodically and symmetrically around a light source. If you observe closely, you will see that the recurring colours are multiple images of the light source. Unlike the halogen light, which produces the colours of the rainbow, the multiple images of a red LED remain red. This happens for refraction too, which shows the close relation between refraction and diffraction.

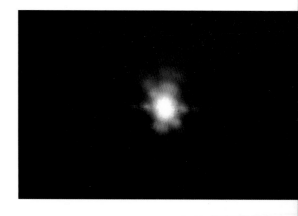

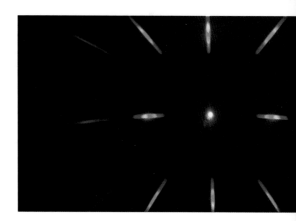

Figure 3.54: The coloured diffraction patterns that arise from viewing a halogen light source through two different kinds of fabric (top and middle) and a grating with narrow parallel lines (bottom).

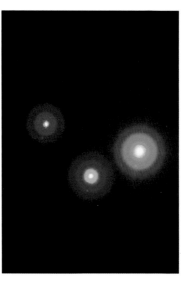

Figure 3.55: A halogen light (left) and three coloured LEDs (right) viewed through lycopodium powder.

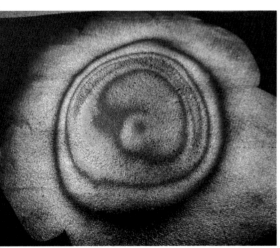

Interference colours

If you dip a ring into soap solution, a thin film will form when you take it out. If you blow into the ring, multicoloured bubbles will appear and float off. If you fill a container with water and let a drop of clear nail polish fall onto the surface, the nail polish will spread out and form a thin layer. If you dip a piece of black card into the water, it will also appear multicoloured when you take it out. Thin transparent films such as oil on water also give rise to colours, as does glass with a thin coating.

Colours that arise on thin films are called 'interference colours' and arise due to multiple reflections between the internal surfaces of a thin transparent medium. Interference colours arise most clearly on layers that are about a micrometre thick. If they are thicker, the colours will contract, lose their saturation and eventually vanish in a diffuse colourless light.

The periodic nature of diffraction and interference colours – the fact that they repeat at regular intervals – was an important early indication of the periodic nature of light. Observations of these phenomena formed the basis of the wave theory developed by the Dutch physicist Christiaan Huygens (1629–95) to account for their periodicity. His wave theory soon displaced the Newtonian corpuscular theory, which conceived of light as small particles called 'corpuscles' and could not explain these periodic colours.

Figure 3.56: A thin film created by a drop of clear nail polish on water is transferred to a piece of black card (top). The thin film on the card makes it appear coloured (bottom).

Figure 3.57 (right): Glass coated with an interference film (sometimes called a 'dichroic film'). When illuminated from the side, the transmitted image (yellow) has the complementary colour to the reflected image (blue).

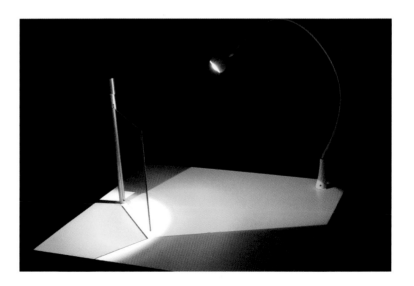

You can investigate the colours of soap bubbles by illuminating a thin film of soap solution and projecting the image onto a wall. In this experiment, a film of soap solution is illuminated with a bright light that is reflected horizontally from below and projected onto the wall using a lens. The large projection allows you to explore the colours and fine movements of the film. To begin with, it is too thick for colours to arise and is fully transparent and colourless like a glass plate. As the liquid runs off and the film becomes thinner, colours arise. Narrow, unsaturated bands of colour appear which then become increasingly saturated and wider as the film gets thinner before vanishing when it bursts. The colours can be used to measure the thickness of the film.

If a film of soap solution is illuminated from the side, light is partially reflected by the film onto the wall, thereby creating a reflected image opposite a transmitted image. The reflected image can be as bright as the transmitted image, allowing you to compare the two. You will notice that the colours of the transmitted image are complementary to the colours of the reflected image.

Figure 3.58: A film of soap solution illuminated by a bright light that is reflected horizontally from below and projected onto a wall using a lens.

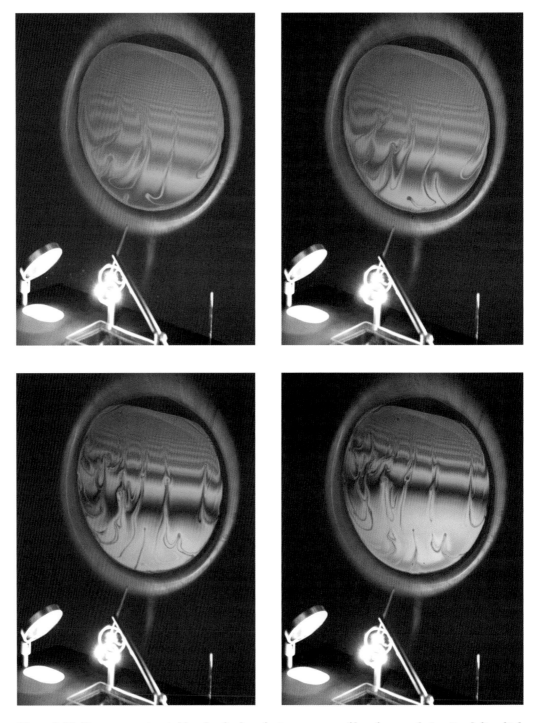

Figure 3.59: Narrow, unsaturated bands of colour first appear on a film of soap solution (top left), which then become more saturated and wider as the film gets thinner (left to right) before vanishing as it bursts (bottom right).

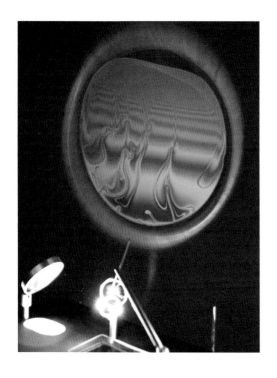

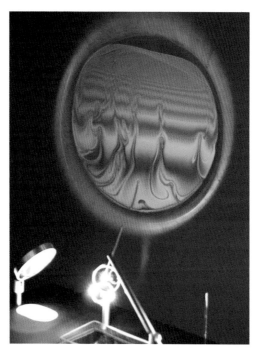

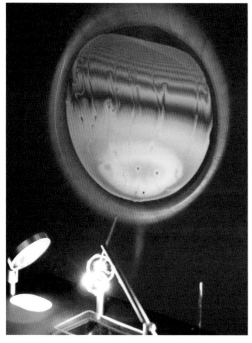

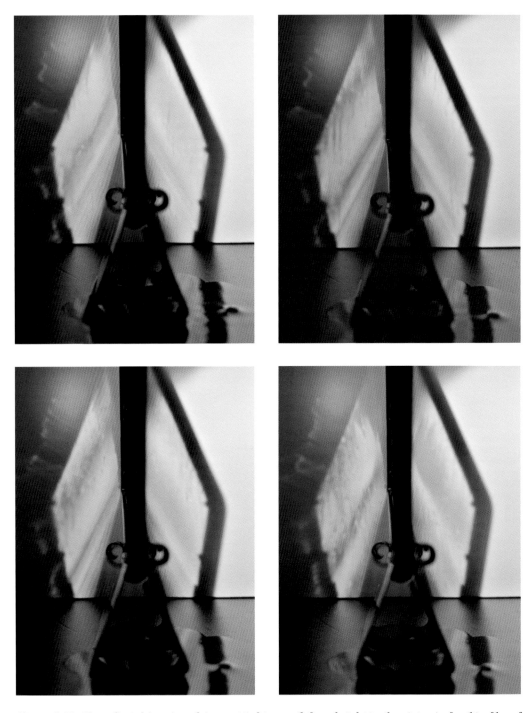

Figure 3.60: The reflected image and transmitted image (left and right in the picture) of a thin film of soap solution (in the centre of the picture) illuminated from the left. The colours in the reflected image are complementary to those in the transmitted image. The coloured bands move further apart as the soap film gets thinner with time (top left to bottom right).

Newton's rings

Newton's rings are coloured fringes that appear between two glass plates clamped together with a thin layer of air in between. It's hard to see the colours simply by looking through the glass plates at a light source, but you can see them well by shining a light source through the glass plates onto a wall. This creates a transmitted image, but, as with the thin film of soap solution, you can shine the light at such an angle that a reflection appears on the wall next to the transmitted image. The colours of the transmitted and reflected images are again complementary to each other.

As the gap gets smaller, the colours move apart until they disappear when it reaches about 0.2 micrometres. As the gap gets larger, the coloured fringes overlap and form a repeating series from magenta to green before disappearing when it reaches about 2 micrometres. Like colours on films of soap solution, Newton's rings belong to the interference colours that arise on thin layers, but the thin layer is now the air. You can also notice such interference colours in cracks in crystals or ice.

With polarisation colours we saw that partial reflection from transparent media creates two images – one image is reflected, the other transmitted. Here again with interference colours we have transparent media creating two images by partial reflection that have a polar relation with respect to colour as well. In this chapter we have seen many examples of colours arising from colourless transparent media, and it is perhaps interference colours that show most clearly that colours arise in complementary pairs.

Figure 3.61: Two glass plates are clamped together with a thin layer of air between them. Illuminated from the right, the colours in the transmitted image are complementary to those in the reflected image (left and right in the picture).

Figure 4.1: The Noah window at Chartres Cathedral.

4. Colours Arising From Coloured Media

In the previous chapter we saw that colours arising from transparent media depend greatly on the geometrical configuration of the elements involved. These phenomena are therefore well suited to being described mathematically. Already in ancient Greece, mathematical descriptions of reflection, refraction and colour were being developed in the nascent science of optics (the science of vision). Despite the central role of these phenomena, optics was primarily concerned with light rather than colour. It was not until Newton's novel experiments described in the next chapter that colour became a central topic in optics and thus part of the mathematical tradition.

Traditionally, the science of colour belonged to chemistry, which is why Goethe refers to surface colours as 'chemical colours'. This discipline was historically separate from the mathematical tradition, being instead part of the experimental tradition that has its roots in medieval medical and alchemical practices. Practitioners within this tradition were generally less interested in developing mathematical hypotheses for understanding phenomena, valuing instead experimental skill and practical knowledge in the manipulation of phenomena. An important goal of experimentation was the production of pigments and dyes. The medieval stained-glass windows are a testament to the impressive skills of early practitioners.

Surface colours

The last chapter demonstrated a plethora of colour phenomena that the sky presents us with, but the earth is no less abundant. From the resplendent tinctures of crystals to the delicate hues of flowers, from the bright markings of insects and birds to the iridescent colours of

Figure 4.2: The chemical colours of iron oxide (top) and copper acetate (bottom), which form on the surface of iron and copper respectively when they react with their environment.

pearls, terrestrial life is teeming with colour. Many colours of natural objects arise because their surfaces contain pigments, such as the oxides that form on metal surfaces when they react with their environment. However, colours such as the feathers of parrots are not just pigments, but pigments embedded in fine structures. Colours such as the iridescent blue wings of morpho butterflies and the intense blue of the marble berry are due solely to fine structures. Thus many of the phenomena that belong to coloured surfaces are closely related to colours arising from colourless media. Nevertheless, surface colours are qualitatively very different from the purer colours of the rainbow and the deep blue of the sky and are bound up with the qualities of the surface and its illumination.

Pigments

Pigments are insoluble colourants. Historically, mineral substances were used, with earth paints being a common pigment since prehistoric times. Red was a common, naturally occurring pigment and sources included cinnabar, minium (also known as red lead), hematite and different kinds of ochre. More rare were saturated green and blue pigments, such as lapis lazuli, which has been used since antiquity because of its intense ultramarine-blue colour and was extremely valuable.

Beginning with the eighteenth century, pigments were produced artificially, such as Prussian blue. Along with inorganic compounds, many synthetically created organic pigments are in use today. The pigments are usually available in the form of fine powders and can be applied and fixed using a binder.

Figure 4.3: Mineral pigments (from top left to bottom right): orpiment, azurite, azurite malachite, cobaltoan calcite, malachite, yellow ochre, red ochre and cinnabar.

Dyes

Dyes are soluble colourants, which is why they have been used for colouring textiles since prehistoric times. Historically, plant and animal dyes were used, such as Tyrian purple made of sea snails (muricidae). With the rise of the chemical industry their usage declined immensely, but they are experiencing a revival nowadays due to their special qualities.

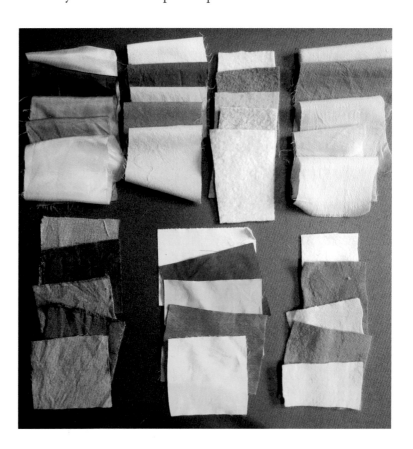

Figure 4.4: Silk, cotton, thick wool, thin wool (top row, left to right), linen, cotton and synthetic felt (bottom row, left to right) dyed for equal amounts of time with cyan, magenta, yellow and black batik dyes. The dyed textiles differ considerably in colour saturation and gloss.

Plant dyes are extracted from roots, leaves, blossoms, seeds and bark and are bound to inorganic salts for application. Ancient cultures had a preference for lightfast dyes from plants such as rubia tinctorum (common madder), named after its red dye, and indigofera tinctoria (true indigo), whose dye is named after it.

*Figure 4.5 (left): Cochineal dye made from insects (**middle**) and plant dyes (clockwise from left): woad, sandalwood, dyer's rocket, ground logwood, rubia tinctorum (common madder) and cornflower.*

Figure 4.6: Wool dyed with cochineal (left), ground logwood (bottom left) and dyer's rocket (below). The wool samples have been dyed in the same tub and fixed with potassium alum.

Figure 4.7: The eye of a peacock's tail feather becomes colourless if the black pigment behind the fine structure is removed.

Structural colours

Not only pigments and dyes give rise to surface colours. There are many materials whose colour depends on fine structures and therefore arise due to diffraction and interference. However, these structural colours often only appear when combined with pigments. A peacock's tail feather, for example, will become almost colourless if the black pigment behind the structures is removed. Conversely, colours arising from thin layers become more intense against a reflective background, as can be observed on the surface of tempered metals. Mother of pearl has thin layers that give it its iridescent colours. As these coloured surfaces are free from pigments, they belong, on the one hand, to colours arising from colourless media. On the other hand, they are bound to a particular object by being a property of the surface and thus also belong to surface colours.

Figure 4.8 (bottom right): Mother of pearl has thin layers which give it its iridescent colours.

Figure 4.9 (below): The tempering colours of a layer of oxide that forms on a shiny metal surface when heated. Its thickness, and therefore colour, depends on the temperature reached before cooling down.

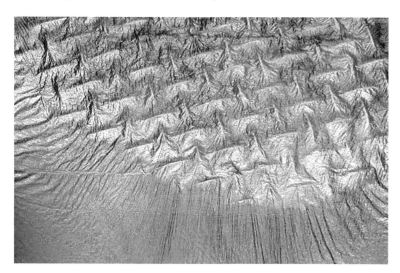

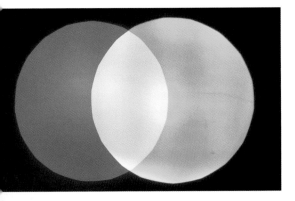

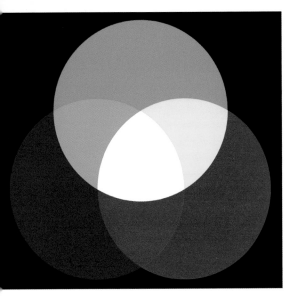

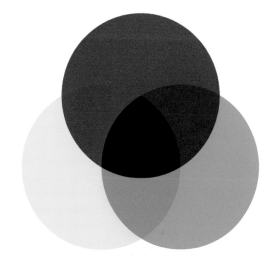

Colour mixing

We saw in Chapter 2 that two complementary colours cancel each other out to produce grey when mixed with a spinning disc. We called this cancelling of complementary colours a basic law of colour mixing. Yet there are other ways of mixing complementary colours. For example, you can shine two lights onto a white screen and place a coloured gel in front of one light and a gel of the complementary colour in front of the other. If the intensities of the two lights are perfectly balanced, cancellation will occur. But rather than producing grey, where the colours overlap you will see white. You can also mix complementary colours by placing a coloured gel in front of a single light and placing a second gel of the complementary colour over the first gel. If the colours of the complementary gels are perfectly matched, where they overlap you will see black.

So while this basic law holds for different kinds of mixing, the cancellation expresses itself in different ways. The overlapping of complementary coloured lights, which is called additive mixing, produces white. The overlapping of complementary coloured gels, which is called subtractive mixing, produces black. However, if the two colours are not complementary, cancellation will not occur. If you additively mix red and blue, two dark colours, you will get magenta, a light colour. Conversely, if you subtractively mix yellow and magenta, two light colours, you will get red, a dark colour. Two dark colours mixed additively produce a light colour, and two light colours mixed subtractively produce a dark colour.

As we have also seen in Chapter 2, more than two colours cancel each other out if all the colours bar one mix to give the complementary of the remaining colour. Colour cancellation will therefore occur when red and blue are mixed additively with green, the complementary colour to magenta. In additive mixing, red, blue and green are the primary dark colours and mixing any two of these produces a secondary light colour, which will be complementary to the third primary colour. Conversely when yellow and magenta are mixed subtractively with cyan, the complementary to red, cancellation again occurs. In subtractive mixing, then, yellow, magenta and cyan are the primary light colours and mixing any two of these produces a secondary dark colour, which will be complementary to the third primary colour. These colour relations can be represented with three overlapping circles.

Mixing transparent paints by painting one layer over another once it has dried is another example of subtractive mixing. However, mixing two opaque pigments on the palette produces a colour whose lightness is between that of the original two colours, such as mixing blue and yellow to produce green. This is a third kind of mixing and belongs to colours arising from the eye. If a colour contrast is seen from far enough away, the separate colours will mix. Newton noticed that if you produce a green pigment by mixing yellow and blue powder and then look at it under a microscope, you can still see the separate yellow and blue particles. The mixing therefore happens optically, in the eye. Similarly, colours on a moving contrast can also mix optically, as we have seen in Chapter 2. Optical mixing is therefore additive with respect to hue because it is like shining two coloured lights onto a screen, but averaging with respect to lightness because the coloured particles are not sources of light, but reflect light.

There is another way to mix colours with the eye. As we have also seen in Chapter 2, if you look at a coloured surface, an afterimage will appear when you look at a white surface afterwards. However, if you look at a coloured surface afterwards, the colour of the surface will mix with the colour of the afterimage. There is relatively little research into this kind of mixing, but unlike optical mixing it is a kind of subtractive mixing.

Figure 4.12: Georges Seurat, The Circus *(1891), Musée d'Orsay, Paris. The colours of the painting are produced by lots of small points of colour on the canvas that mix optically when you look at it from a distance.*

Figure 4.10 (top left): Lights of two complementary colours projected onto a screen mix additively to produce white.

Figure 4.11: In additive mixing (middle left) any two of the three dark primary colours mix to produce a light secondary colour, which is the complement to the third primary colour. The three primary colours mix to produce white. In subtractive mixing (bottom left) any two of the three light primary colours mix to produce a dark secondary colour, which is the complement of the third primary colour. The three primary colours mix to produce black.

Additive mixing with coloured lights

Three light sources, such as torches or projectors, and coloured gels can be used to explore additive mixing.

If the lights are not coloured, each on its own appears white on a white wall. If two of them partially overlap, only the area where they overlap will appear white and the other illuminated areas will appear darkened, or grey. If all three partially overlap, only the area where they all overlap will appear white and the other illuminated areas will appear darkened.

If you place a red, blue and green gel in front of the lights, you can mix lights of the additive primary colours with each other. Where two coloured lights overlap, magenta, cyan or yellow will appear. Where all three overlap, white appears.

Normally, red, green and blue gels are used for additive mixing. However, if you use a light blue, a dark blue and a grey gel instead, you do not get shades of blue, but the usual red, green and blue and its mixtures as before, but slightly less saturated. This phenomenon is explored further in Chapter 6.

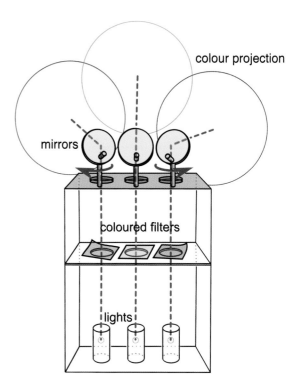

Figure 4.13 (right): Diagram of three light sources projected through coloured gels and reflected by movable mirrors onto a wall.

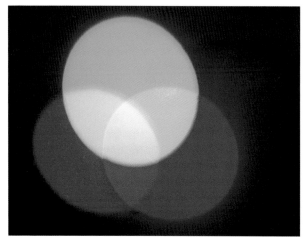

Figure 4.14 (left): Additive colour mixing using red, blue and green gels.

Figure 4.15 (top right): Close-up of the three movable mirrors from the front.

Figure 4.16 (bottom right): Close-up of the overlapping colours on the wall.

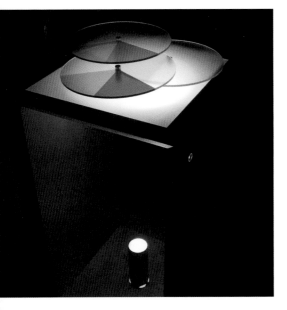

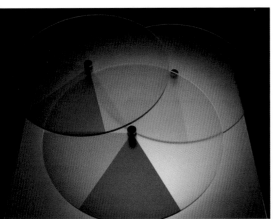

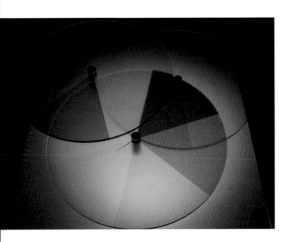

Subtractive mixing with coloured gels

Subtractive colour mixing can be explored by placing coloured gels in front of a single light source. You can produce different colours by placing a yellow, cyan and magenta gel in front of a light source in different combinations. Yellow and cyan together produce green, yellow and magenta produce red, and cyan and magenta produce blue. Cyan, magenta and yellow all together produce grey or black. The darkness of the grey depends on the saturation of the mixed colours.

In another experiment, three rotatable, coloured discs are mounted on an illuminated platform. The colours of the discs are the subtractive primary colours cyan, magenta and yellow. Each disc has five segments with increasing saturation. In the middle area, a segment from each of the three discs overlap, to each side, a segment from two. This allows you to experiment with combining different shades of the three primary colours.

Effects of colour on us

So far we have explored the diverse conditions under which colours arise. We have looked at colours arising from the eye, from transparent media and lastly from coloured media. What we have not touched upon yet is how colour affects us and our lives. Colour is all around us, influencing our feelings, thoughts and even actions. The appearance of the first flowers in spring has an uplifting effect on us and the rainbow awakens a sense of awe and wonder every time we see it.

Goethe was a pioneer in studying the psychological effects of colour, and the didactic part of his *Farbenlehre* describes the effects of individual colours on the human being. By connecting the conditions under which colours arise in nature with the way we experience colours inwardly, he shows that the transition from colours in science to colours in art is not abrupt, but gradual. Although there are undoubtedly differences in approach, there is no hard boundary between the colours of the scientist and the colours of the artist.

The *Red Window* in the Goetheanum in Switzerland is an example of colour phenomena that belong to both science and art. On the one hand, the effect of being immersed in red light is palpable – the beating of our heart changes and we feel a vitality in our body. We can also experience an emotional depth. To these physiological and psychological effects Rudolf Steiner adds depictions of beings that introduce what Goethe calls the 'allegorical, symbolic and mystical' dimensions of colour that concludes the didactic part of the *Farbenlehre*.

Figure 4.19 (above): Rudolf Steiner and Assja Turgenieff, Red Window. *Goetheanum, Switzerland.*

Figure 4.17 (top left): Three rotatable discs of the subtractive primary colours, cyan, magenta and yellow, are mounted on an illuminated platform.

Figure 4.18: (middle and bottom left): Close-up of the three rotatable discs. When the magenta disc is rotated, the colours yellow and magenta (middle) mix to produce red (bottom).

Looking through coloured windows

You can investigate the effects that different colours have on you by placing a large, coloured gel over a window in a room with a single window (looking through a coloured gel will have a similar effect, but not as pronounced). While most of the experiments described in this book allow you to explore the conditions under which colours arise, here you can explore what arises in yourself with colour as the condition.

When you look through a coloured window you will notice that objects change colour. If the colour of the object is similar to the colour of the gel, the object will appear strikingly light, whereas the complementary colour will appear darkened. Green leaves viewed through magenta, for example, will look almost black. As with subtractive mixing, complementary colours are darkened. If you look through a coloured window for some time its colour decreases and the complementary colour can begin to reappear. The green leaves that originally looked almost black through a magenta window will begin to look green again. If you then remove the gel and look through the window, the environment will appear in the complementary colour to the gel.

This process of adaptation has been described in connection with coloured shadows in Chapter 2, **which** arise when the eye adapts to the colour of the illumination by adding the complementary colour. Here, the eye adapts to the colour of the gel by adding the complementary colour, which becomes visible when you look away. However, what you can experience clearly for the first time is the effect of being immersed in a colour. If you direct your attention inwards to your own experiences that arise as a response to the colour, you will notice the different moods that everything assumes when seen through different colours.

Figure 4.20 (top): View of a yellow-, a magenta- and a cyan-coloured window. The yellow window is displaced by the water prism.

Figure 4.21 (bottom): A magenta- and a blue-coloured window. The tetrahedral water prism reflects the magenta window.

Figure 4.22 (top): The sun shining through a magenta-coloured window.

Figure 4.23 (bottom): The sun shining through a yellow-and-blue coloured window.

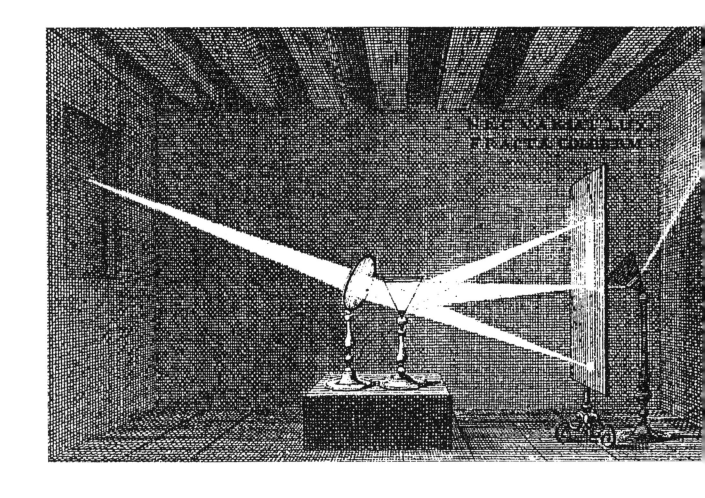

Figure 5.1: A version of Newton's experimentum crucis *with a lens illustrated in the second French edition of the* Opticks *(Traité d'Optique, 1722).*

5. The Complementarity of Colour Experiments

In the previous two chapters we investigated two kinds of colours: colours that arise from colourless media, such as raindrops, and colours that arise from coloured media, such as coloured glass. Investigations into these two kinds of colours have a separate history prior to Newton; the former belonged to the mathematical tradition, the latter to the experimental tradition. Newton was one of the first scientists to bring these two traditions together with his *Opticks*, first published in 1704. Almost all of the experiments in the *Opticks* were conducted in a darkened room or camera obscura. Newton famously darkened his room and made a small hole in the window shutter to allow a narrow beam of sunlight to shine in. Even today, the vast majority of scientific experiments on light and colour are conducted in darkened rooms in the Newtonian tradition.

Goethe criticised Newton for experimenting almost exclusively in a darkened room, insisting that the study of light and colour must also include outdoor observations in nature. From his experiments looking through a prism, he realised that colours only arise at a contrast between light and dark, and that light and darkness therefore enjoy an equal status in such experiments. However, Goethe discovered that this does not only hold for prism experiments, but for colour experiments in general. His discovery of the complementarity of colour experiments has wide-ranging implications that are still being explored today. In this chapter we will consider some examples of this complementarity.

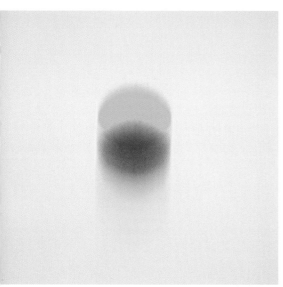

Figure 5.2: Spectrum of a light sun in a dark universe (top) and a dark sun in a light universe (bottom).

Colour experiments and their inversion

Goethe realised that not only experiments looking through a prism can be inverted by exchanging black and white, but also prism experiments with a light source and a screen. While acknowledging the value of experiments in dark environments, he considered that limiting the diversity of colour phenomena to such experiments was experimentally unjustified. In the polemical part of his *Farbenlehre* he attempted to invert many of Newton's experiments by exchanging the role of light and darkness.

Light and dark sun through a prism

Newton was the first person to propose that sunlight consists of coloured components. Having darkened his room, he made a hole in the window shutter so that a small white image of the sun appeared on the floor. When he placed a prism in the beam of light, the solar image appeared as an elongated coloured spectrum on the opposite wall. Newton concluded that the previously white solar image was actually composed of the many coloured solar images on the wall. The prism did not produce the colours but merely separated or 'analysed' the light of the sun by refracting different colours to different positions on the wall.

Newton's experiment can be repeated by shining a light through a small circular aperture in an opaque plate and reflecting it onto the wall with a mirror. Placing a prism between the mirror and the wall produces a colourful, elongated image such as the one Newton saw. Newton presents this experiment in his *Opticks* as part of his experimental proof that 'the light of the sun consists of rays differently refrangible'.

Goethe was not convinced by Newton's argument. He had discovered that spectra occur not only with light sources, such as the sun, but also with dark shadows in light surroundings. What would Newton's spectrum look like if the sun were dark in a dazzling, light universe?

This can be demonstrated by substituting a glass plate with a small circular shadow caster in the centre for the opaque plate with an aperture. Now, instead of being dark, the wall is illuminated with light, and instead of the familiar Newtonian spectrum, you can see an inverted spectrum of colours – each of the colours in the Newtonian spectrum has been replaced with its complementary colour.

By applying Newton's argumentation to this inverted experiment, we could explain it by saying that 'the darkness of the dark sun consists of rays differently refrangible'. Although Goethe pointed this out, he did not believe in a 'theory of darkness' that was complementary to Newton's theory of light. Instead, he claimed that if you are convinced of Newton's proposition, you should also be convinced of its inversion because it is based on the same logic. He took the inverted statement to be absurd and thus rejected the original statement too.

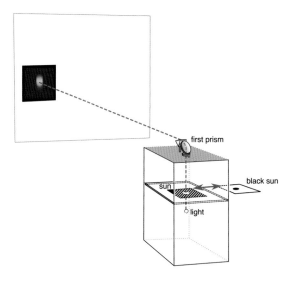

Figure 5.3: Diagram of the optical arrangement. The miniature white sun in a dark universe can be replaced by a miniature dark sun in a light universe (horizontal red arrow).

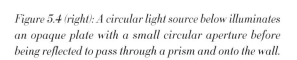

Figure 5.4 (right): A circular light source below illuminates an opaque plate with a small circular aperture before being reflected to pass through a prism and onto the wall.

Light and dark sun through crossed prisms

To prove his new theory of light further, Newton analysed the coloured spectrum created by the first prism using a second prism. He placed the second prism in the path of the coloured light coming from the first prism, but oriented perpendicular to it – the two prisms were crossed. The same coloured spectrum appeared on the wall as it did the first time, only now it was oriented diagonally instead of vertically. Since no new colours appeared and the existing colours had merely shifted position while maintaining the same order, Newton proposed that every colour has its own unique displacement, or degree of refraction.

Assuming that the first prism works the same way as the second one, we can trace the path from each colour on the wall backwards to the light source. We could then say that in Newton's experiment, 'the coloured rays come from the light sun'. This train of thought supports Newton's second proposition that all the spectral colours are contained in sunlight.

Goethe, however, thought this proposition was inadequate for several reasons. In particular, he believed that all of Newton's experiments could be inverted: a dark sun in a dazzling, light universe also generates a spectrum – one that is complementary in colour to Newton's. Because of the technical difficulties involved, Goethe could not **invert Newton's crossed prism** experiment. It was first described in the late 1950s by a colour research group led by Norwegian writer and poet André Bjerke (1918–85) at Oslo University.

If Newton's crossed prism experiment is repeated using the previous apparatus and placing the second prism between the first and the wall, you will see that the spectrum is merely displaced and oriented diagonally. However, if you invert this experiment in exactly the same way as in the previous experiment, you will see that the analysed spectrum of the dark sun behaves exactly the same as it did for Newton's spectrum, that is each of the complementary colours has its own unique degree of refraction. As with the previous experiment, we could claim that it proves the inverse of Newton's claim. Here we could say that the 'coloured rays come from the dark sun'.

So is darkness made up of coloured components? This was definitely not Goethe's opinion. Rather, he felt that the symmetry of the experiments left him with no choice but to conclude that colour arises from the interplay of light and darkness.

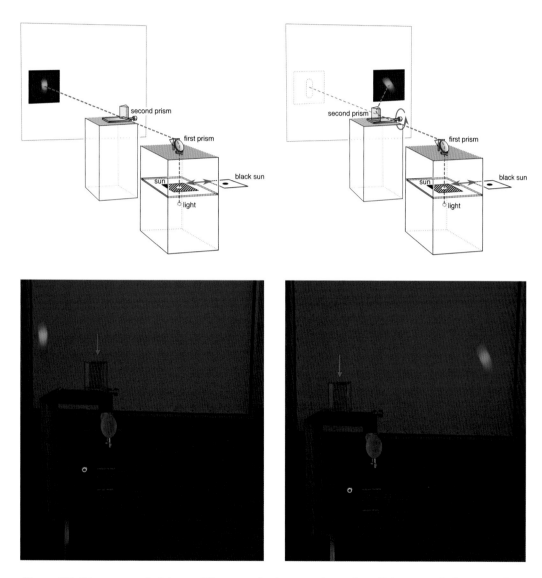

Figure 5.5: Diagrams and pictures of the crossed prism experiment for a light sun in dark surroundings before (left) and after (right) the second prism is added. The light sun can be replaced by a dark sun indicated by the horizontal red arrows in the diagrams. When the second prism, indicated by the vertical arrow in the pictures, is placed between the first prism and the vertical spectrum, the spectrum is displaced to the right and becomes diagonally orientated.

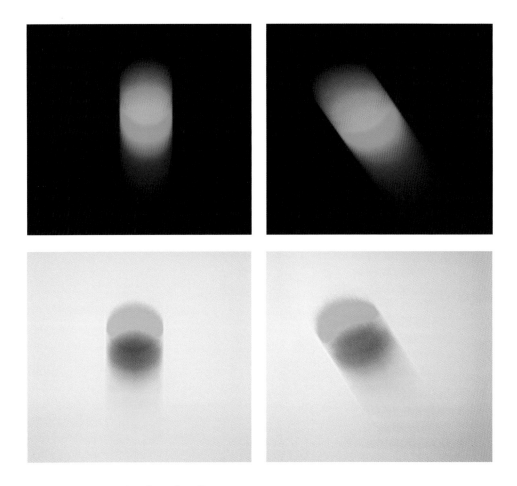

Light shadows

We have seen that if we lived in a light universe with a dark sun, Newton would have set about proving propositions about darkness instead of light. What other differences would we observe in this universe in which light and darkness are inverted? What about shadows? Would they not be light instead of dark?

By moving a dark object in front of a long, thin light source, such as your hand in front of a strip light, you can investigate how an object casts a light shadow when a 'source of darkness' is present. If you place an object between a strip light and the wall, there will be a diffuse dark shadow on the wall. If you move your hand in front of the strip light, you will notice a 'light shadow' that is lighter than the surrounding area and that moves in the opposite direction to your hand.

Figure 5.6 (above): The spectrum of a light sun in a dark universe (top pair) and the spectrum of a dark sun in a light universe (bottom pair). The right-hand picture in each pair shows the respective spectrum after the second prism is added.

You can use a complicated shape such as a rectangular frame as a shadow caster. If you look carefully, you will see a light shadow of the frame appear surrounded by a darkened area as you put your hand in front of the strip light. This phenomenon was first discovered by German physicist Georg Maier (1933–2016).

The key to understanding this phenomenon is to first understand a dark shadow and then invert the relationship between light and darkness. We know that a dark shadow is cast by an object obstructing the light. But where does the shadow's darkness come from? When, during a total solar eclipse, we enter into the full shadow cast by the moon (the umbra), we see only the darkness of space. The darkness of the moon's shadow cast onto the earth is due to this surrounding darkness.

This is the case for all shadows cast by objects obscuring the light: the shadow's darkness is due to the darkness of the environment. However, although the shadow of the moon is completely dark during a total solar eclipse, our shadow on a sunny day is not. This is because the sky on a clear day is not completely dark either and our shadow is illuminated by light from the sky. In other words, the darker the environment, the darker the shadow.

Figure 5.7 (left): A rectangular frame casts a light shadow when 'illuminated' by a dark object in a light surrounding.

Figure 5.8 (above): A dark sun in a light universe simulated by an object in front of a strip light.

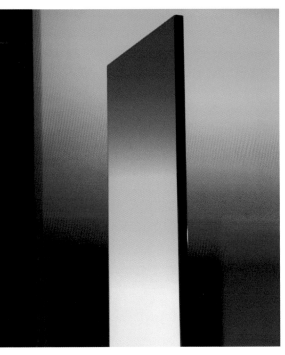

Figure 5.9 (top and bottom): Prismatic colours reflected in a mirror in front of their complementary prismatic colours projected onto a screen.

Now imagine again that we are in the moon's shadow during a total solar eclipse. If the surrounding space were to suddenly light up, we would no longer be in a dark shadow, but illuminated from all directions. The only darkness we would see would be the darkness of the moon obscuring the sun. And if the sun were to suddenly cease shining, we would not notice at first because it is behind the moon. Only when the dark sun emerged would we see a second dark object against the lightness of space. This second dark object would also darken the area where we are standing – the dark sun would act as a source of darkness. We have just moved out of the light shadow cast by the moon. The same happens when we see a light shadow in the experiments above: the shadow's lightness is due to the lightness of the environment and our hand darkens the area surrounding the shadow.

For a shadow to appear there must be a contrast between a source and the surrounding environment. Only in the case of the light shadow, a dark object in a light environment acts as a source of darkness and the shadow's lightness is due to the lightness that surrounds a dark object.

Colour experiments and their generalisation

Newton and Goethe are often regarded as antagonists with opposing worldviews. In the previous experiments we have indeed emphasised differences in their experimental approach. Yet what is often missed is that despite their differences they also share important similarities. Both Newton and Goethe were interested in the physiology, physics and even metaphysics of colour, both conducted research in all these areas, and, most importantly of all, both were empiricists who believed that fundamental principles can be found by an analysis of phenomena.

So far in this chapter an experiment from Newton has been followed by an inverted 'complementary experiment' from Goethe. But as we have seen in previous chapters, complementary colour phenomena can arise together. With interference colours in Chapter 3 the phenomenon of partial reflection was used to create two images, a reflected image and a transmitted image. Newton's rings and the colours on a thin film of soap solution can be seen with a reflected image and a transmitted image next to each other. This juxtaposition showed that the reflected image has the complementary colours of the transmitted image.

Newton's and Goethe's experimental innovations can be combined in just this way. We can take an optical innovation from Newton and generalise it to include Goethe's experimental innovation of showing the complementarity of light and darkness together. As with interference colours, a reflective surface in the experimental setup creates two inverse, mutually dependent images. A mirror thus allows an experiment and its counterpart to appear simultaneously. Indeed, they are unified in a single experiment. The innovation of using a mirror to combine two complementary experiments into a single experiment goes back to Norwegian physicist Torger Holtsmark (1924–2014).

This approach can be extended to other colour experiments, such as the additive mixing of coloured lights that we saw in the previous chapter. Such experiments allow new insights into the symmetry of colour phenomena and the underlying conditions from which they arise. As we will see, these experiments are not special cases of colour phenomena, but rather the generalisation of earlier special cases that showed only half of a phenomenon. Usually, the other half is darkened and rendered invisible by absorbing the light. Replacing a black surface with a mirror minimises this absorption and allows two partial phenomena to appear simultaneously as two aspects of a single, unified phenomenon.

Generalising colour experiments by replacing black surfaces that absorb light with mirrors conserves the total amount of the light's energy in the experiment, rather than allowing some of it to be converted into thermal energy by absorption. Thus, this mutual arising of a colour phenomenon and its complement results from the conservation of energy, a law of physics discovered after Goethe's time.

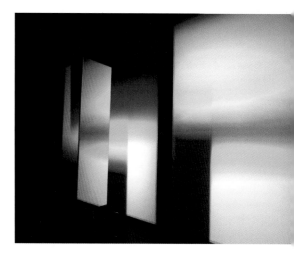

Figure 5.10: The screen on the left behind the mirror shows the two edge spectra coming together to create the familiar Newtonian spectrum. The screen on the right shows the complementary spectrum formed when the edge spectra are brought together in the opposite way. The mirror on the left reflects the complementary spectrum on the right such that the reflection is complementary to the Newtonian spectrum behind it.

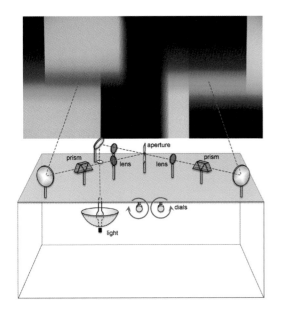

Figure 5.11: Diagram of the optical arrangement with the projected complementary spectra in the background. The gap between the two halves of the aperture as well as their vertical alignment can be adjusted by turning the dials at the front.

Figure 5.12: Complementary spectra on a screen with the optical setup in the foreground.

Complementary spectra with a mirror aperture

In Chapter 3 we saw that the blues of the sky and the oranges and reds of the setting sun arise together in mutually determined, complementary pairs. The following experiment demonstrates the same symmetry in the production of spectral colours. Whenever a given spectral colour is produced in this experiment, its complementary colour always appears simultaneously.

The leading role in this experiment is played by the aperture at the centre of the optical arrangement. It consists of two small mirrors positioned one above the other with a small gap between them. It is illuminated from the left by a powerful light source. Instead of having to look at the mirror aperture first from one side and then from the other, both sides are reflected onto a large screen in front and focused using two lenses. Just as the lens of the eye focuses the image on the retina, each of the lenses focuses an image of one side of the aperture onto the screen. The image on the left side of the screen corresponds to the view of the aperture from the left, the image on the right side from the right.

Since the aperture is illuminated from the left, when viewed from the left its reflective surface will appear light against the dark background. Conversely, viewed from the right, the aperture blocks the light and therefore appears dark against a light background. The mirror aperture thus brings about its own inversion of light and darkness, producing two images of itself that are mutually dependent. By placing prisms after the lenses the two mutually dependent images of the aperture seen through a prism are projected onto the screen as two mutually dependent complementary spectra.

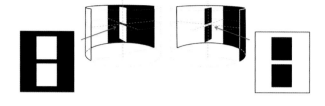

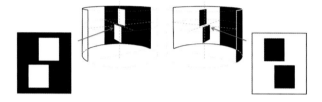

Figure 5.13: The mirror aperture is made up of two mirrors positioned one above the other with a light surrounding on one side and a dark surrounding on the other. Viewed from the light surrounding (left arrow) the mirror aperture appears light against the dark surrounding. Viewed from the dark surrounding (right arrow) the mirror aperture appears dark against the light surrounding. The two halves of the mirror aperture can be aligned vertically (top diagram) or offset horizontally (bottom diagram).

Figure 5.14: Two views of the optical arrangement from in front of the screen looking at the reflected light source when standing in the green region (left) and the complementary magenta region (right). Because Newton's experiment is inverted using a mirror aperture, the magenta light is not produced by mixing the spectral components red and blue from the two ends of the Newtonian spectrum, but is produced fully analagous to the green light by exchanging light and darkness.

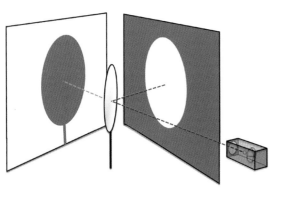

Figure 5.15: Diagram of the basic principle behind additive and subtractive colour mixing with three coloured light sources and a mirror. Only one light source is shown here, which simultaneously creates a shadow in light surroundings (grey dashed line) and a light reflection (red dashed line) in dark surroundings.

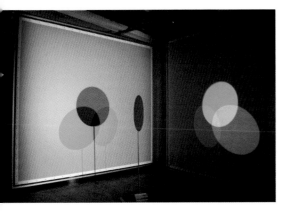

Additive and subtractive colour mixing with a mirror

We saw in the previous chapter that we could mix coloured lights additively, and coloured transparent media subtractively. We also saw that there was an inverse relation between these two kinds of mixing: additive mixing always produces a lighter colour, subtractive mixing a darker one. We can understand the relation between these two kinds of mixing with an experiment that uses an elliptical mirror illuminated with coloured lights.

When illuminated by coloured lights, an oblique elliptical mirror simultaneously casts circular coloured shadows on one screen and circular coloured lights on another. By adjusting the position of the mirror, the circles can be made to overlap. The coloured lights mix additively to produce white, while the coloured shadows mix subtractively to produce black.

This demonstrates that the two processes of colour mixing are mutually dependent – each is an inversion of the other. Colour mixing is therefore not subtractive because it is based on coloured media, but rather because it starts from light surroundings and mixes colour by darkening. Additive mixing, on the other hand, starts from dark surroundings and mixes colour by lightening. In dark surroundings, colours appear lighter than their surroundings and therefore act as lights. The mixed colour is lighter than the two original colours. Conversely, in light surroundings colours appear darker than their surroundings and therefore act as shadows. The mixed colour is darker than the two original colours.

Figure 5.16: An elliptical mirror illuminated by a red, a blue and a green light. The coloured lights are reflected by the mirror onto the dark screen on the right, and mix additively to produce a white screen on the left. The magenta shadow is cast by the green light, the cyan shadow by the red light, and the yellow shadow by the blue light. Where coloured lights overlap on the right-hand screen, they mix additively. Where coloured shadows overlap on the left-hand screen they mix subtractively. The amount of overlapping and mixing of the coloured lights and shadows can be changed by moving the mirror.

Figure 5.17: The different views of the left-hand screen when walking around the mirror, which remains in the same position (from top left). The image seen in the mirror is a reflection of the right-hand screen and is the exact complement of the image on the screen behind it.

Prismatic colours and colour mixing

In the previous chapter we saw that coloured lights mix additively and coloured media mix subtractively. However, the previous experiment shows that coloured media mix subtractively not simply because they are coloured media, but because coloured media mix by darkening the light. Once we realise that subtractive mixing is not restricted to transparent media, we can apply it to cases of mixing by darkening where no coloured media are involved. The previous experiment is one example, prismatic colours are another.

We saw in the section on prisms in Chapter 3 that when we bring two edge spectra together, two colours overlap to form a third colour. When we bring yellow and cyan together, they overlap to produce green. We also saw that when we bring the two edge spectra together in this way, the spectrum is darkened. If we compare this with the previous experiment, we see the same principle: two light colours, yellow and cyan, mix by darkening to produce a darker colour, green. With magenta arising by mixing blue and red we have the inverse situation: two dark colours, blue and red, mix by lightening to produce a lighter colour, magenta.

What is remarkable is that this explanation in terms of additive and subtractive mixing works in the opposite direction too. We can start with the end situation, which is the three colours red, green and blue in the first spectrum. As we move the prism closer, we are lightening the spectrum. Where red and green overlap, yellow appears, and where green and blue overlap, cyan appears. This is again just as we saw with additive mixing in the previous experiment. Conversely, we can start with the end situation of the other spectrum, which is the three colours yellow, magenta and cyan. As we move the prism closer, we are darkening the spectrum. Where yellow and cyan overlap, green appears, and where cyan and magenta overlap, blue appears. This accords with subtractive mixing as we saw it in the previous experiment.

Newton explained the structure of the spectrum in terms of components of light that are refracted by different amounts.

This account thus explains the order of colours in the spectrum in terms of properties of light, not of colour. Goethe's approach to understanding the structure of the Newtonian spectrum was to think of it in terms of colour mixing. Although Goethe did not develop the idea beyond the arising of green and magenta out of the edge spectra, we can see that this idea can be developed with the idea of additive and subtractive mixing to provide an understanding of the structure of both Newton's spectrum and its complementary spectrum.

Figure 5.18: Two screens are placed in front of three coloured lights on the left-hand side of the room. The lights illuminate an elliptical mirror, which reflects the lights onto the left-hand screen, where additive mixing occurs, and casts coloured shadows onto the right-hand screen, where subtractive mixing occurs. A spectrum and its counterpart are produced using a mirror aperture and projected onto two screens on the right-hand side of the room. The right-hand spectrum and the right-hand mixing of coloured lights show green arising from the subtractive mixture of yellow and cyan. The left-hand spectrum and the left-hand mixing of coloured lights show magenta arising from the additive mixture of blue and red.

Figure 6.1: Claude Monet drew the same haystacks through the changing light of the day and the changing seasons.

6. Sight and Coloured Light

We can learn a lot about colour vision just by observing coloured objects in nature. For example, carefully observe a banana in the afternoon sun. Although we can see that the surface away from the sun is darker and that the shape of the banana creates surfaces of different shades, we do not see a mosaic of individual patches of different colours. Rather, we see a three-dimensional object of a single colour in an illuminated environment.

In order for the artist to reproduce how an object appears in an illuminated environment, the artist needs to unlearn this normal way of seeing and learn to see each surface as a different colour. If the sun begins to set as we are observing our banana, the colour of the banana does not change, but the colour of the illumination cast by the sun. Claude Monet (1846–1926) would draw the same object again and again to capture how different illumination and atmospheric conditions affect how the object appeared.

The colour of an object remains constant in changing illumination because the eye adapts to the illumination. The process of adaptation occurs continuously without our noticing it. However, adaptation has its limits. By investigating phenomena at this limit, we can understand colour vision. We have already looked at colours arising from the process of adaptation when we discussed coloured shadows in Chapter 2. We will now investigate further the effects of coloured light for sight.

Colour and illumination

We saw in Chapter 2 that the eye adapts to the colour of the illumination by adding the complementary colour, thus ensuring that the colour of objects remains constant, but producing coloured shadows in the process. However, adaptation fails in cases of strongly coloured illumination, for example in the case of bright, coloured spotlights in stage lighting. When this happens, we no longer see the surface colour of an object, but the colour of the light reflecting from it.

You can experiment with this phenomenon using two coloured gels. Hold a yellow gel in your outstretched hand and place a cyan gel in front of your eye with your other hand. The yellow gel will still appear yellow. Now move the cyan gel towards the yellow gel. The yellow gel will turn green. Although you are looking through the cyan and yellow gels in both situations, in the first the eye adapts to the colour of the cyan gel as if it were the colour of the illumination, and so the yellow gel still appears yellow. But when you move the cyan gel towards the yellow gel there comes a point when the eye can no longer adapt to the colour of the cyan gel. In this situation, the colour of the object, the yellow gel, appears coloured by the cyan gel and thus turns green. We now have the normal case of subtractive colour mixing, which we have seen in Chapter 4.

Similarly, with strongly coloured illumination we no longer see the surface colour of an object, but the colour of the illumination. As with the yellow gel, however, if the object is coloured, the colour of the illumination is affected by the colour of the object. We will see examples of this in the following sections.

Figure 6.2: A yellow gel seen through a cyan gel close to the eye looks yellow (top). If the cyan gel is moved away from the eye so that it is in front of the yellow gel, the combination now looks green (bottom). In both cases we have subtractive colour mixing because we are looking through two coloured gels. In the first case, however, the eye adapts to the colour of the cyan gel as if it were the colour of the illumination by adding the complementary colour and cancelling out the effect of the cyan gel.

Surface colour and illumination

The experiment with coloured gels illustrates that for coloured lights we perceive the colour of illumination to the degree to which the surface colours disappear and vice versa. We will now consider situations in which it is not so much the colour of the illumination that we cannot fully adapt to, but rather the inability of certain kinds of illumination to render surface colours visible. Looking at such cases helps us better understand the role of light in seeing coloured objects.

You can investigate the relation between surface colour and illumination by viewing a coloured painting illuminated first with natural light, then a tungsten light, and finally an LED. As you change the light source, the painting changes dramatically in its expression and richness of colour. Some colours become extremely intense, while others become surprisingly grey or even black.

By comparing the different light sources in this way, you can appreciate the importance of illumination for a rich colour experience. We saw in Chapter 1 that light makes sight possible. Here we can see to what degree different kinds of light make it possible to see coloured objects.

Figure 6.3: A Mondrianesque picture illuminated by natural light (top), a tungsten lightbulb (middle) and an LED (bottom).

Monochromatic and polychromatic illumination

Now view a coloured painting illuminated with a halogen lamp and a sodium lamp. If you place a yellow gel in front of the halogen lamp so that both light sources look yellow, you will hardly be able to tell the difference between them when looking at a white wall. With a coloured painting on the wall, however, you will be surprised at how much it changes. When illuminated by the halogen lamp, the colours remain more or less as they did with daylight illumination. When illuminated by a sodium lamp, the colours turn to shades of yellow and grey. If you observe carefully, you will notice that blue turns particularly dark.

As you can see many colours in the illumination from the halogen lamp, it can be characterised as 'polychromatic illumination'. As you can only see a single colour in illumination from the sodium lamp, it can be characterised as 'monochromatic illumination'. These two terms do not refer to the colour of the light sources; for they both look yellow. Rather, they refer to the colours that can be seen when the painting is illuminated by the light source.

Figure 6.4 (top): Coloured squares illuminated by a halogen lamp (above) and a sodium lamp (below).

Figure 6.5 (left): Wax crayons illuminated by a halogen lamp (above) and a sodium lamp (below).

Figure 6.6: Paintings on a wall illuminated by a sodium lamp (top left); by a sodium lamp on the left and a halogen lamp with a yellow gel on the right (top right and bottom left); and only a halogen lamp with a yellow gel (bottom right).

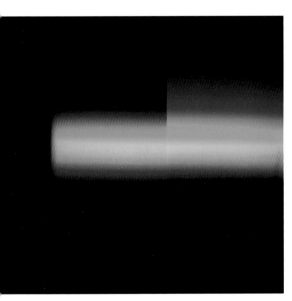

Figure 6.7: Yellow card on a white background illuminated by daylight (top). When placed in a spectrum produced by a prism, the yellow card behind the left side of the spectrum darkens the blue part of the spectrum (bottom).

Surface colour and spectral illumination

Illumination is monochromatic when a multicoloured object appears only in the colour of the illumination. If you shine the light from a prism onto a coloured painting, you will find that this light is also monochromatic. If you illuminate a coloured painting with the yellow portion of the spectrum, the situation is just like the sodium lamp: the surface appears as shades of yellow and grey. The relationship between monochromatic illumination and surface colour can be investigated systematically by placing different coloured surfaces, such as coloured cards, in the spectrum created by projecting light through a prism.

When we look at a coloured surface illuminated by the whole spectrum, we notice that a particular band or bands of colour are darkened more than others. For example, a blue surface will darken the yellow portion of the spectrum. Conversely, a yellow surface will darken the blue portion of the spectrum. A magenta surface will darken the green portion of the spectrum, whereas a green surface will darken both the red and the violet parts of the spectrum; the two ends of the spectrum that mix additively to create magenta. Here again we find complementarity: a coloured surface will darken light in the spectrum that is its complementary colour. The colour that an object appears, then, is complementary to the colour of the light that it absorbs.

The selective darkening of spectral colours by coloured objects shows that objects of different colours darken the light in different ways. When we see the colour of an object, what we see is the way in which the surface of the object darkens the illumination. This phenomenon shows the relation between the prismatic colours in Chapter 3 and surface colours in Chapter 4. It is also the basis of spectroscopy, an extremely useful analytical tool in physics and chemistry.

Figure 6.8: A range of coloured cards illuminated by a white light (top left). In the cyan spectral band the red cards are darkened (top right), in the green spectral band the magenta cards are darkened (bottom left) and in the yellow spectral band the blue cards are darkened (bottom right).

Multicolour projection

When we look at a landscape we can usually distinguish many nuances of colour. These are reduced to shades of grey in a black-and-white photograph. The Scottish physicist James Clerk Maxwell (1831–79) was the first person to develop a procedure for reproducing scenes in colour. Maxwell took three black-and-white photographs of a tartan ribbon through a red, a green and then a blue gel. Using three projectors, he projected each of the three black-and-white images through the gel through which they were photographed. When the three-coloured images were aligned, a single-coloured image appeared which, due to additive mixing, reproduced most of the different hues and nuances of colour.

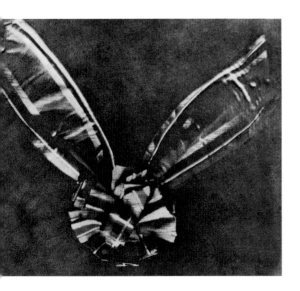

Figure 6.9: Maxwell's tartan ribbon (1861). Photograph of a projection showing Maxwell's invention of using black-and-white photographs taken with coloured gels to reproduce colour.

Three-colour projection

The American inventor of the Polaroid camera, Edwin H. Land (1909–91), repeated Maxwell's experimentation by projecting three black-and-white images through the coloured gels through which they were photographed.

Land also experimented with removing one of the coloured gels from in front of a projector. To his surprise, most of the colours remained. We have already encountered this **phenomenon in Chapter 4 when we** explored additive mixing with coloured lights. As we saw, when the red gel is replaced with a grey one and the green gel is replaced with a light-blue one, a good reproduction of the range of colours is still possible. Colour mixing using a colourless, light-blue and dark-blue gel to produce the colours of additive mixing cannot be explained through additive colour mixing alone.

Figure 6.10: A colour image of a still life (top left) is produced by overlaying the projections of a black-and-white image taken and projected through a blue, a green and a red gel.

Figure 6.11: A colour image (top left) of Vincent van Gogh's painting Fishing Boats on the Beach at Saintes-Maries, produced by projecting the top right black-and-white photograph through a blue gel, the bottom left photograph through a green gel and the bottom right photograph through a red gel and then aligning the three images on a screen. If the red gel is replaced with a grey gel and the green gel with a light-blue gel, rather than the image appearing as shades of blue, the reds, greens and the yellows (the additive mixture of red and green) will remain.

Two-colour projection

When carrying out his experiments with three black-and-white slides and three coloured gels, Land also discovered that you can see more colours than you would expect if you project only two slides and use only a red gel. You can see more than just shades of red between light and dark.

English colour researcher Michael Wilson (1901–85) systematised these experiments with his double projection experiment. In this experiment, two light and dark contrast patterns are projected onto a screen. The black, white and shades of grey overlap in manifold ways on the wall. If one of the slides is projected through a red gel, one does not just see shades of red, but also shades of green and blue.

This experiment is closely related to coloured shadows. Whereas coloured shadows are normally cast by an opaque object, here semi-transparent media cast partial shadows. As Wilson was familiar with Goethe's colour experiments, he was able to show that Land's discovery was not a new phenomenon, but a striking example of coloured shadows, which we saw in Chapter 2. In both three-colour and two-colour projection experiments, adaptation balances the colour of illumination by adding the complementary colour. What Wilson's experiment shows is that when the eye perceives colours, it perceives the way that the coloured surface darkens the illumination. Here we find again that when we see colour, we are seeing the way in which the colours of objects darken the light.

Figure 6.12: A colourless, a light-blue and a dark-blue gel with a cross located at a different position on each are placed in front of three light sources and projected onto the wall. The yellow cross is the shadow cast by the dark-blue light, the blue cross by the light-blue light and the red cross by the colourless light.

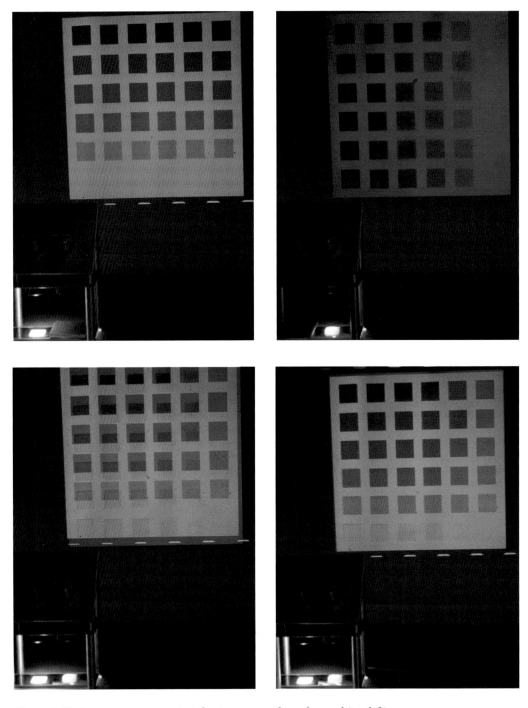

Figure 6.13: Grey squares are projected onto a screen through no gel (top left) and through a red gel (top right). When they are projected together colours start to appear (bottom left) and when the grey squares are aligned a whole range of colours appears (bottom right).

Figure 6.14: Shades of grey projected onto a screen through no gel (top left) and through a red gel (top right). When they are projected together colours appear (left).

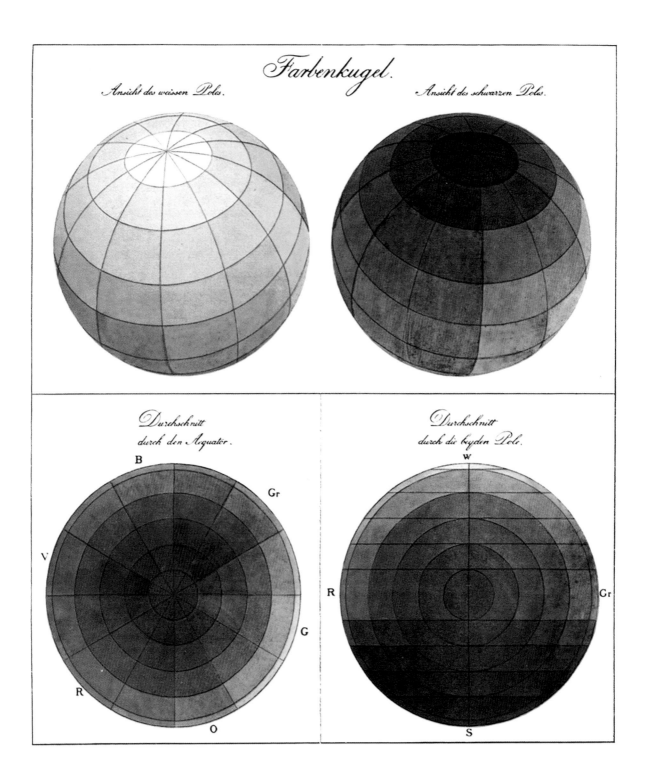

Figure 7.1: Philipp Otto Runge's colour sphere (1810).

7. The Order of Colours

We have seen in the preceding chapters that while nature might at times present us with an isolated colour, this colour is intimately related to other colours. Sometimes a simple observation can reveal these relations. We saw in Chapter 2 that looking at a colour calls forth a complementary afterimage. In Chapter 3 we saw that the setting sun presents us with a sequence of colours from yellow to red. Sometimes, however, experimental apparatus is required to show the relation between an observed colour and other colours. In Chapter 5, for example, we needed to replace a dark surface with a mirror to reveal colours that are otherwise suppressed.

All the colour phenomena we have seen express in their own unique way two fundamental principles about the appearance of colour.

The first fundamental principle is that each colour has one complementary colour, which we could call the 'principle of complementarity'.

This seems like a physiological phenomenon, for we first saw this principle expressed in colours arising from the eye. But in later chapters we saw that prismatic colours and colour mixing also express this principle. It is initially puzzling how two completely different processes – one physiological, one physical – can give rise to the same lawfulness. It is surprising, for example, that inverting the physical illumination, as the mirror does with complementary spectra and colour mixing in Chapter 5, results in complementary colours.

However, the complementary colour isn't exactly the same colour in the different processes. If you observe these two classes of colour phenomena closely, you will notice slight deviations between physiological and physical complementary pairs of colours. For example, if we mix yellow and blue lights to create a colourless

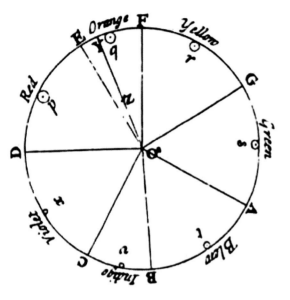

Figure 7.2: The colour circle from Newton's Opticks.

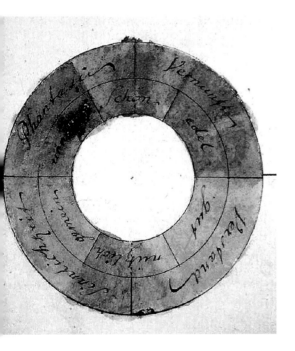

Figure: 7.3: Goethe's colour circle.

mixture, the afterimage of the yellow light will be more violet than the blue light. Nevertheless, the general principle holds as it finds expression in the different processes. We could say that complementarity is a general law of colour that is modified by the process through which it is expressed.

The second fundamental principle is that each colour has many neighbouring hues, which we could call the 'principle of continuity'.

Again, this seems like a physiological phenomenon: surely it is due to the nature of the eye that yellow can gradually change into orange, and orange into red. Yet this colour sequence is displayed by the sun as its position in the sky gradually changes as it approaches the horizon. Here, a continuous change in the physical conditions brings about a continuous change in the observed colour.

Different representations of colour have been developed to represent these two fundamental principles.

Colour circles

Aristotle represented colours on a line between white and black based on the lightness of each colour. This linear representation of colour was dominant until the eighteenth century. One of the first two-dimensional representations of colour was Newton's colour circle.

Newton's colour circle

Newton realised that a colour circle can be used as a device for determining the result of mixing two spectral colours and designed his colour circle with this function in mind. Even though he did not distinguish between additive and subtractive mixing, his rules for mixing were a crucial breakthrough for additive colour mixing and the development of colour solids, described below.

The colour circle in Newton's *Opticks* is formed by bending the solar spectrum into a circle so that the red and violet ends touch. Newton divided the solar spectrum into seven colours based on the proportions of the seven tones of the Dorian musical scale, which corresponds roughly to the observed colours.

Newton's colour circle represents the second of the two fundamental principles mentioned above, but not the first.

Goethe's colour circle

Unlike Newton, Goethe did not base his colour circle on a single phenomenon but developed his colour circle to represent the many different colour phenomena we have seen. In particular, he wanted to represent the complementarity of colour phenomena and placed complementary colours opposite each other on the circle. Goethe's colour circle therefore includes magenta, which is reduced to a line in Newton's colour circle because it does not appear in the solar spectrum. Goethe's colour circle thus represents both fundamental principles of the appearance of colour.

Goethe was also interested in characterising the relationship between two colours. Three such characterisations are possible based on the geometrical relations of six colours in a circle. The relation between complementary colours, which are diametrically opposed, he characterises as *harmonic*, the relation between neighbouring colours as *characterless*, and the relation between colours connected by two inscribed triangles as *characteristic*.

We have seen in Chapter 5 that there is an intimate relationship between the prismatic colours and colour mixing. Although Goethe was not aware of this, the two triangles representing characteristic colour pairs represent additive and subtractive mixing. At the three corners of the triangle pointing upwards are the three primary colours of subtractive mixing and at the corners of the triangle pointing downwards are the three primary colours of additive mixing.

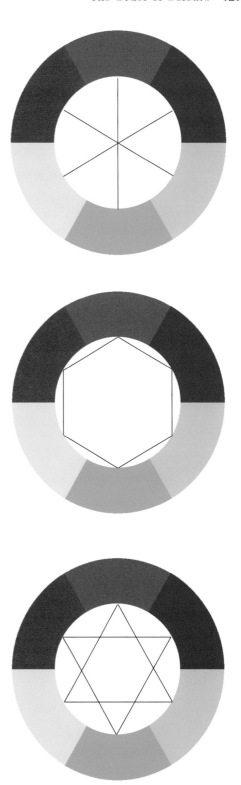

Figure 7.4: Goethe characterised the relationship between complementary pairs of colours, which are diametrically opposed on his colour circle, as 'harmonic' (top), between neighbouring pairs as 'characterless' (middle) and between pairs connected by two inscribed triangles as 'characteristic' (bottom). The upward pointing triangle joins the three primary colours of subtractive mixing, while the downward point triangle joins the three primary colours of additive mixing.

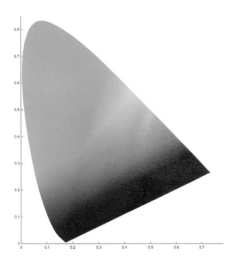

Figure 7.5 (above): The CIE chromaticity diagram. The curved edge represents the colours of the solar spectrum. The colours within the diagram represent mixtures of spectral light and the neutral grey point in the middle of the diagram (also called the white point) lies between any colour and its complementary colour.

The CIE chromaticity diagram

The CIE chromaticity diagram was created by the *Commission Internationale de L'éclairage* (CIE), or the International Commission on Illumination, in 1931 to quantify the correlation between an observed colour and the physical properties of light. Being a two-dimensional representation of colour, the chromaticity diagram can be regarded as a modern colour circle. The curved edge represents the colours of the solar spectrum, but the two ends of the spectrum are separated to include magenta. The colours within the diagram represent mixtures of spectral light and the neutral grey point in the middle of the diagram (also called the white point) lies between any colour and its complementary colour. Because of these requirements, as well as the attempt to represent perceptible differences quantitatively, its shape is no longer circular. The CIE chromaticity diagram adopts principles from both Newton's and Goethe's colour circle.

Colour solids

Colour circles and other plane figures can represent all the possible combinations of mixing saturated colours with each other, but not of adding white or black, which create different shades of the same hue. To include white and black, a third dimension is therefore necessary: a solid figure must be used to represent colour. Such a three-dimensional representation of colour is called a 'colour solid'.

Ordering colours

You can explore the order of colours by ordering coloured objects, such as coloured blocks. You can begin by constructing sequences of neighbouring colours. For example, you can start with yellow and make a sequence through orange to red. Similarly, you can construct sequences from blue to red, and from blue to yellow. You can also build sequences of the same colour but of different shades of lightness, and then develop these sequences into two dimensional arrangements. You can then develop these into three-dimensional arrangements of colours: you can construct a colour solid.

You can also experiment with colour constellations, create colour harmonies or dissonances using similar colours or strong contrasts, and observe how a colour appears when surrounded by a different colour. In this way you can investigate Goethe's colour harmonies. But as with looking through coloured windows as described in Chapter 4, these observations can become introspective. You can observe your own inner experiences of colour, your own sensibility, by exploring how different combinations affect you. Goethe called this aspect of colours the 'sensory-moral effect of colours'.

Figure 7.6: Coloured wooden blocks arranged in three-dimensional patterns to form colour solids.

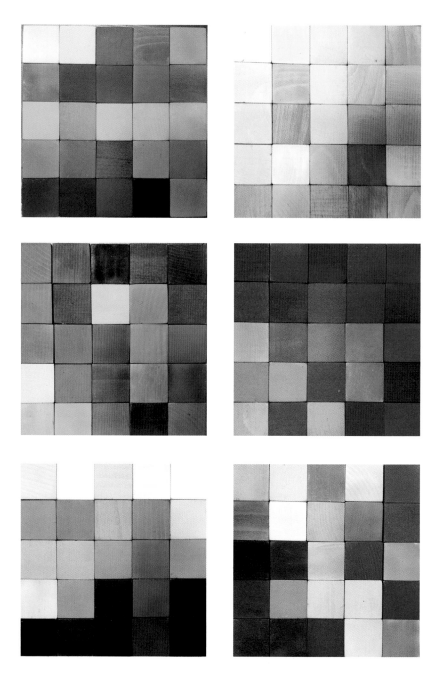

Figure 7.7: Blocks coloured different shades of blue, yellow, green, red, grey, and pinks and purples, with a few blocks of different colours included for comparison. Nuances of different hues in shades of the same colour can be seen, as well as shades that have transitioned to a different colour. Also, as in Chapter 4, each colour theme creates a different mood.

Runge's colour sphere

One of the earliest colour solids was developed by the German painter Philipp Otto Runge (1777–1810), who used a sphere to extend the circle into the third dimension. Runge represented colours on a sphere by placing the saturated colour circle around the equator, with white at the north pole and black at the south.

Munsell's colour solid

Runge's colour sphere is characterised by its symmetry; each basic colour is assigned an area of equal size on the surface. However, if we observe the light hues near the white pole, we will notice that we can distinguish many more light yellows than light blues. Conversely, if we observe the dark hues near the black pole there are many blue hues, but no yellow hues at all, for a darkened yellow becomes brown. If this polarity of yellow and blue is taken into account, the colour solid becomes asymmetrical. The American painter Albert Henry Munsell (1858–1918) developed a colour solid based on the perceptible differences between hues. Like the CIE diagram, as soon as perceptible differences are taken into account, colour solids become asymmetric. But even in this asymmetry we find again the expression of the polarity of colour appearance, namely between yellow as a light colour and blue as a dark colour.

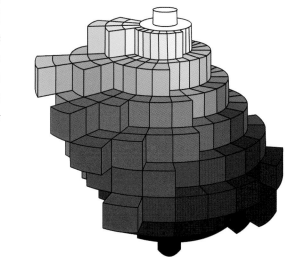

Figure 7.8 (top right): Philipp Otto Runge's colour sphere (1810) showing the white pole (top left) and the black pole (top right), and a cross section through the equator (bottom left) and the poles (bottom right).

Figure 7.9 (bottom right): Munsell's empirical colour solid is asymmetrical because near the white pole there are many more discernible light-yellow hues than light-blue hues, and near the black pole there are many more discernible blue hues and no yellow hues.

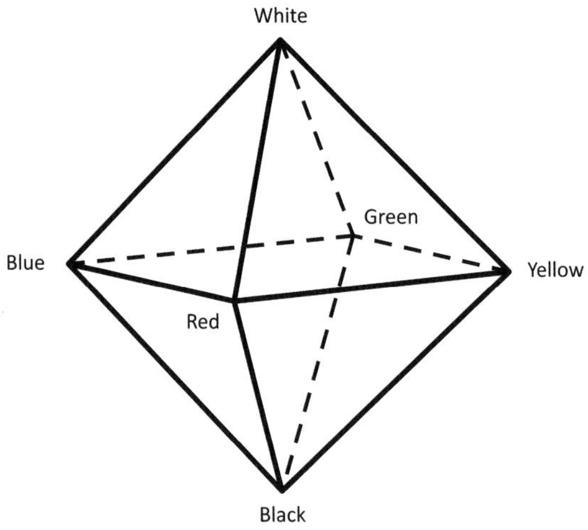

Figure 7.10: Wittgenstein's Colour Octahedron.

Wittgenstein's colour octahedron

The didactic part of Goethe's *Farbenlehre* includes a letter in which Runge claims that reddish green and yellowish blue are impossible colours. The philosopher Ludwig Wittgenstein (1889–1951) wanted to understand why.

Goethe noticed that all colours can be described in terms of the following six words: red, yellow, blue, green, white and black. We can see this with Runge's colour sphere. If we move around the equator starting with yellow, we can describe the colours as increasingly reddish yellows, until finally there is no trace of yellow left. We have a continuous sequence from yellow to red in which the intermediate orange hues can be described as yellowish red. From red we can continue the sequence through bluish red to blue, from blue through greenish blue to green, and finally from green through greenish yellow back to yellow. We can therefore describe all the colours on the equator in terms of four colours. Sequences from white to black also form a continuous sequence through a third colour: we can start from white, go through whitish red to red, and through blackish red to black.

So why can't reddish green or yellowish blue exist? We can express our colour terms using a colour octahedron by putting one of our six colour terms at each corner of an octahedron. When we order the colours in this way, we discover that the impossible colours are opposite each other on the octahedron. We might think that yellowish blue is green because we are used to this combination when mixing paints. But we can find a green that is perfectly balanced between blue and yellow, just as we can find a yellow that is perfectly balanced between blue and red. Wittgenstein believed that this perspicuous representation of our basic colour term shows us why certain colour combinations such as reddish greens or yellowish blues are impossible.

What does a colour circle or solid represent?

These different colour circles and colour solids represent, in different ways, the two principles mentioned at the beginning of this chapter: each colour has one complementary colour and many neighbouring hues. But do they represent anything more?

Newton used his colour circle to calculate what colour is produced when spectral colours are mixed. It is thus a quantitative representation of an underlying physical mechanism that gives rise to colours. We have seen that colours arise in very different ways: from the eye, from colourless media, from coloured media. Goethe discovered that although colour arises through very different processes, there is nevertheless an order in the way colour arises that is independent of these processes. Thus, rather than the colour circle representing an underlying physical mechanism, for Goethe it represents the lawfulness of the appearance of colour itself.

Wittgenstein came to the same conclusion. The colour octahedron is based on the idea of three colour pairs, which was first proposed by the German physiologist Ewald Hering (1834–1918). Hering, however, proposed it together with an underlying mechanism in the eye to explain the oppositions between these pairs of colours. For Hering, then, the colour octahedron represents an underlying physiological mechanism. Wittgenstein realised that an underlying mechanism cannot explain why it is the *colour* blue, and not some other *colour*, that is opposed to yellow. Or why the *colour* white, and not some other *colour*, is opposed to black. For Wittgenstein, the colour octahedron represents the logic of colour – it shows the internal relations between the colours themselves.

Wittgenstein often returned to Goethe's fundamental insight:

> The ultimate goal would be: to grasp that everything in the realm of fact is already theory. The blue of the sky shows us the basic law of chromatics. Let us not seek for something behind the phenomena – they themselves are the theory.[5]

Conclusion

In the Introduction we saw that Goethe did not want to disprove Newton's theory of colour, but see colour in a new way, just as Galileo had seen a swinging lamp in a new way. In the case of Galileo, his new way of seeing had profound consequences, not only for science but for Western culture as a whole – it helped develop a scientific way of seeing that we could call a 'scientific consciousness'. What implications does Goethe's way of seeing colour have for us?

One of the central ideas of Thomas Kuhn's *Structure of Scientific Revolutions*, expressed in the quote that begins the Introduction, is that scientific progress is not merely a process of finding ever better theories to explain a fixed, objective reality. Rather, scientific progress is interrupted by radical changes in how we see the world, by scientific revolutions. This is because science is always based on a particular way of seeing, on a particular paradigm. And at certain junctures, one paradigm gets displaced by another.

A result of paradigm change, and a reason why it occurs, is that particular phenomena that remained stubbornly inexplicable in the old paradigm – 'anomalies' as Kuhn called them – can be understood in the new paradigm. As Kuhn stresses, this new understanding is not because the new paradigm simply provides a better theory for the same phenomena, but rather because the phenomena themselves have been transformed by the new paradigm. By seeing the motion of the swinging church lamp as essentially reaching its initial position after each swing rather than reaching its natural resting place after a long period of time, Galileo was able to demonstrate that many disparate kinematic phenomena reveal an inner unity and lawfulness expressed through laws of motion.

The question, then, is what new understanding of colour Goethe's way of seeing can bring. Newton saw colours as belonging purely to the light – as an effect of the ray's 'refrangibility', or wavelength as we would say today. He saw darkness, on the other hand, as the mere absence of light; darkness, for Newton, plays no role in the arising of colour. Goethe's original *aperçu* was that colour only appears at a boundary between light and darkness. He saw colour as belonging neither to the light, nor to the darkness, but as an effect of their interaction. This way of seeing allowed the two principles mentioned in Chapter 7, namely the complementarity and continuity of colour phenomena, to express a lawfulness of colour itself, not of an underlying physical of physiological mechanism.

The consequences of this are illustrated well by the contrast drawn in Chapter 5 between Newton's and Goethe's explanation of the order of the colours of the rainbow. For Newton, the order of the rainbow's colours is due to the 'diverse refrangibility' of the sun's rays. The order of the *colours themselves*, however, is purely arbitrary. That the *colour* green should appear at the centre of the rainbow and not some other *colour* is, for Newton, completely inexplicable. For Goethe, green appears because it is the subtractive mixture of yellow and cyan. But this explanation is not available to Newton because no *physical mixing of light* occurs. This is not a problem for Goethe. For him subtractive mixing is an expression of a law of colour, not of light. By relating colours to a property of light, namely its 'refrangibility', Newton removed the possibility for an explanation in terms of relations between the colours themselves. As a result, in the received view, which is based on Newton's theory, it must forever remain a surprising coincidence that green just happens to be the subtractive mixture of yellow and cyan.

Goethe can explain the order of the other colours in the rainbow, too. We saw in Chapter 3 that the blue sky and setting sun express two continuous colour sequences, starting with white at one end and ending in darkness at the other. As we also saw, Goethe was able to show that these two colour sequences appear at a boundary between light and dark viewed through a prism. Since the rainbow appears when the sun is reflected and refracted by multiple raindrops, each of the two edges of the sun creates one of each sequence, which overlap in the middle because of the small angular size of the sun, producing green.

By demonstrating that the order of the colours of the rainbow expresses the principles of complementarity and continuity, Goethe is able to explain this order by showing that the lawfulness at work in other colour phenomena is at work here too. Just as Galileo's laws of motion express the inner unity of diverse kinematic phenomena, the principles of complementarity and continuity express the inner unity of diverse colour phenomena.

It is important to stress that Newton's way of seeing can provide an *explanation* of the order of the colours of the rainbow, just not in terms of colours, but rather in terms of underlying invisible causes. The colours themselves, for Newton, are inherently isolated, bearing no direct relation to each other. As a result, this way of seeing will bring us no closer to the essence of colour, colour itself must forever remain a mystery to it. Goethe's way of seeing, on the other hand, can provide an *understanding* of colour. But only if we learn to see colours as Goethe did. This means coming to see the many expressions of complementary and continuity presented in this book as expressions of the essence of colour itself.

This, in itself, is reason enough for us to follow Goethe on this journey. But there is also a larger possibility inherent in coming to see colours as Goethe did. Newton's way of seeing colours is largely due to Galileo's scientific paradigm. For while Galileo thought that the geometrical properties of objects, including their motion, properly belonged to the world, he also thought that colours belonged only to the human mind. This is the deeper reason why Newton wanted to explain colours in terms of geometrical properties of light – he wanted an explanation of colour in terms of entities that are independent of the human mind.

The dualistic scientific paradigm that separates mind and world shapes how we see the world today – certainly in the West and increasingly in the rest of the world, too. We are educated to see colours as Newton did, so it can take enormous effort to see them as Goethe did. But if this journey through Goethe's world of colour has helped us to do so, then we have taken a step towards a new kind of consciousness – a consciousness for which mind and world are not separate entities, but two sides of the same coin.

The Experiment As Mediator Between Object and Subject

J.W. von Goethe

As the human being becomes aware of objects in his environment he will relate them to himself, and rightly so since his fate hinges on whether these objects please or displease him, attract or repel him, help or harm him. This natural way of seeing and judging things seems as easy as it is essential, although it can lead to a thousand errors – often the source of humiliation and bitterness in our life.

A far more difficult task arises when a person's thirst for knowledge kindles in him a desire to view nature's objects in their own right and in relation to one another. On the one hand he loses the yardstick which came to his aid when he looked at things from the human standpoint; i.e., in relation to himself. This yardstick of pleasure and displeasure, attraction and repulsion, help and harm, he must now renounce absolutely; as a neutral, seemingly godlike being he must seek out and examine what is, not what pleases. Thus the true botanist must remain unmoved by beauty or utility in a plant; he must explore its formation, its relation to other plants. Like the sun which draws forth every plant and shines on all, he must look upon each plant with the same quiet gaze; he must find the measure for what he learns, the data for judgement, not in himself but in the sphere of what he observes.

The history of science teaches us how difficult this renunciation is for man. The second part of our short essay will discuss how he thus arrives (and must arrive) at hypotheses, theories, systems, any of the modes of perception which help in our effort to grasp the infinite; the first part of the essay will deal with how man sets about recognising the forces of nature. Recently I have been studying the history of physics and this point arose frequently – hence the present brief discourse, an attempt to outline in general how the study of nature has been helped or hindered by the work of able scientists.

We may look at an object in its own context and the context of other objects, while refraining from any immediate response of desire or dislike. The calm exercise of our powers of attention will quickly lead us to a rather clear concept of the object, its parts, and its relationships; the more we pursue this study, discovering further relations among things, the more we will exercise our innate gift of observation. Those who understand how to apply this knowledge to their own affairs in a practical way are rightly deemed clever. It is not hard for any well-organised person, moderate by nature or force of circumstance, to be clever, for life corrects us at every step. But if the observer is called upon to apply this keen power of judgement to exploring the hidden relationships in nature, if he is to find his own way in a world where he is seemingly alone, if he is to avoid hasty conclusions and keep a steady eye on the goal while noting every helpful or harmful circumstance along the way, if he must be his own sharpest critic where no one else can test his work with ease, if he must question himself continually even when most enthusiastic – it is easy to see how harsh these demands are and how little hope there is of seeing them fully satisfied in ourselves or others. Yet these difficulties, this hypothetical impossibility, must not deter us from doing what we can. At any rate, our best approach is to recall how able men have advanced the sciences, and to be candid about the false paths down which they have strayed, only to be followed by numerous disciples, often for centuries, until later empirical evidence could bring researchers back to the right road.

It is undeniable that in the science now under discussion, as in every human enterprise, empirical evidence carries (and should carry) the greatest weight. Neither can we deny the high and seemingly creative independent power found in the inner faculties through which the evidence is grasped, collected, ordered, and developed. But how to gather and use empirical evidence, how to develop and apply our powers – this is not so generally recognised or appreciated.

We might well be surprised how many people are capable of sharp observation in the strictest sense of the word. When we draw their attention to objects, we will discover that such people enjoy making observations, and show great skill at it. Since taking up my study of light and colour I have often had opportunity to appreciate this. Now and then I discuss my current interests with people unacquainted with the subject: once their attention is awakened they frequently make quick note of phenomena I was unaware of or had neglected to observe. Thus they may be able to correct ideas developed in haste, and even produce a breakthrough by transcending the inhibitions in which exacting research often traps us.

Thus what applies in so many other human enterprises is also true here: the interest of many focused on a single point can produce excellent results. Here it becomes obvious that the researcher will meet his downfall if he has any feeling of envy which seeks to deprive others of the discoverer's laurels, any overwhelming desire to deal alone and arbitrarily with a discovery.

I have always found the cooperative method of working satisfactory, and I intend to continue with it. I am aware of the debts I have incurred along the way, and it will give me great pleasure later to acknowledge these publicly.

If man's natural talent for observation can be of such help to us, how much more effective must it be when trained observers work hand in hand. In and of itself, a science is sufficient to support the work of many people, although no one person can carry an entire science. We may note that knowledge, like contained but living water, rises gradually to a certain level, and that the greatest discoveries are made not so much by men as by the age; important advances are often made by two or more skilled thinkers at the same time. We have already found that we owe much to the community and our friends; now we discover our debt to the world and the age we live in. In neither case can we appreciate fully enough our need for communication, assistance, admonition, and contradiction to hold us to the right path and help us along it.

Thus in scientific matters we must do the reverse of what is done in art. An artist should never present a work to the public before it is finished because it is difficult for others to advise or help him with its production. Once it is finished, however, he must consider criticism or praise, take it to heart, make it a part of his own experience, and thereby develop and prepare himself for new works. In science, on the other hand, it is useful to publish every bit of empirical evidence, even every conjecture; indeed, no scientific edifice should be built until the plan and materials of its structure have been widely known, judged and sifted.

I will now turn to a point deserving of attention; namely, the method which enables us to work most effectively and surely.

When we intentionally reproduce empirical evidence found by earlier researchers, contemporaries, or ourselves, when we re-create natural or artificial phenomena, we speak of this as an experiment.

The main value of an experiment lies in the fact that, simple or compound, it can be reproduced at any time given the requisite preparations, apparatus, and skill. After assembling the necessary materials we may perform the experiment as often as we wish. We will rightly marvel at human ingenuity when we consider even briefly the variety of arrangements and instruments invented for this purpose. In fact, we can note that such instruments are still being invented daily.

As worthwhile as each individual experiment may be, it receives its real value only when united or combined with other experiments. However, to unite or combine just two somewhat similar experiments calls for more rigour and care than even the sharpest observer usually expects of himself. Two phenomena may be related, but not nearly so closely as we think. Although one experiment seems to follow from another, an extensive series of experiments might be required to put the two into an order actually conforming to nature.

Thus we can never be too careful in our efforts to avoid drawing hasty conclusions from experiments or using them directly as proof to bear out some theory. For here at this pass, this transition from empirical evidence to judgement, cognition to application, all the inner enemies of man lie in wait: imagination, which sweeps him away on its wings before he knows his feet have left the ground; impatience; haste; self-satisfaction; rigidity; formalistic thought; prejudice; ease; frivolity; fickleness – this whole throng and its retinue. Here they lie in ambush and surprise not only the active observer but also the contemplative one who appears safe from all passion.

I will present a paradox of sorts as a way of alerting the reader to this danger, far greater and closer at hand than we might think. 1 would venture to say that we cannot prove anything by one experiment or even several experiments together, that nothing is more dangerous than the desire to prove some thesis directly through experiments, that the greatest errors have arisen just where the dangers and shortcomings in this method have been overlooked. I will explain this assertion more clearly lest I merely seem intent on raising a host of doubts. Every piece of empirical evidence we find, every experiment in which this evidence is repeated, really represents just one part of what we know. Through frequent repetition we attain certainty about this isolated piece of knowledge. We may be aware of two pieces of empirical evidence in the same area; although closely related, they may seem even more so, for we will tend to view them as more connected than they really are. This is an inherent part of man's nature; the history of human understanding offers thousands of examples of this, and I myself make this error almost daily.

This mistake is associated with another which often lies at its root. Man takes more pleasure in the idea than in the thing; or rather, man takes pleasure in a thing only insofar as he has an idea of it. The thing must fit his character, and no matter how exalted his way of thinking, no matter how refined, it often remains just a way of thinking, an attempt to bring several objects into an intelligible relationship which, strictly speaking, they do not have. Thus the tendency to hypotheses, theories, terminologies, and systems, a tendency altogether understandable since it springs by necessity from the organisation of our being.

Every piece of empirical evidence, every experiment, must be viewed as isolated, yet the human faculty of thought forcibly strives to unite all external objects known to it. It is easy to see the risk we run when we try to connect a single bit of evidence with an idea already formed, or use individual experiments to prove some relationship not fully perceptible to the senses but expressed through the creative power of the mind.

Such efforts generally give rise to theories and systems which are a tribute to their author's intelligence. But with undue applause or protracted support they soon begin to hinder and harm the very progress of the human mind they had earlier assisted.

We often find that the more limited the data, the more artful a gifted thinker will become. As though to assert his sovereignty he chooses a few agreeable favourites from the limited number of facts and skilfully marshals the rest so they never contradict him directly. Finally he is able to confuse, entangle, or push aside the opposing facts and reduce the whole to something more like the court of a despot than a freely constituted republic.

So deserving a man will not lack admirers and disciples who study this fabric of thought historically, praise it, and seek to think as much like their master as possible. Often such a doctrine becomes so widespread that anyone bold enough to doubt it would be considered brash and impertinent. Only in later centuries would anyone venture to approach such a holy relic, apply common sense to the subject, and – taking a lighter view – apply to the founder of the sect what a wag once said of a renowned scientist: 'He would have been a great man if only he hadn't invented so much.'

It is not enough to note this danger and warn against it. We need to declare our own views by showing how we ourselves would hope to avoid this pitfall, or by telling what we know of how some predecessor avoided it.

Earlier I stated my belief that the direct use of an experiment to prove some hypothesis is detrimental; this implies that I consider its indirect use beneficial. Here we have a pivotal point, one requiring clarification.

Nothing happens in living nature that does not bear some relation to the whole. The empirical evidence may seem quite isolated, we may view our experiments as mere isolated facts, but this is not to say that they are, in fact, isolated. The question is: how can we find the connection between these phenomena, these events?

Earlier we found those thinkers most prone to error who seek to incorporate an isolated fact directly into their thinking and judgement. By contrast, we will find that the greatest accomplishments come from those who never tire in exploring and working out every possible aspect and modification of every bit of empirical evidence, every experiment.

It would require a second essay to describe how our intellect can help us with this task; here we will merely indicate the following. All things in nature, especially the commoner forces and elements, work incessantly upon one another; we can say that each phenomenon is connected with countless others just as we can say that a point of light floating in space sends its rays in all directions. Thus when we have done an experiment of this type, found this or that piece of empirical evidence, we can never be careful enough in studying what lies next to it or derives directly from it. This investigation should concern us more than the discovery of what is related to it. To follow every single experiment through its variations is the real task of the scientific researcher. His duty is precisely the opposite of what we expect from the author who writes to entertain. The latter will bore his readers if he does not leave something to the imagination, while the former must always work as if he wished to leave nothing for his successors to do. Of course, the disproportion between our intellect and the nature of things will soon remind us that no one has gifts enough to exhaust the study of any subject.

In the first two parts of my *Contributions to Optics* I sought to set up a series of contiguous experiments derived from one another in this way. Studied thoroughly and understood as a whole, these experiments could even be thought of as representing a single experiment, a single piece of empirical evidence explored in its most manifold variations.

Such a piece of empirical evidence, composed of many others, is clearly of a higher sort. It shows the general formula, so to speak, that overarches an array of individual arithmetic sums. In my view, it is the task of the scientific researcher to work toward empirical evidence of this higher sort – and the example of the best men in the field supports this view. From the mathematician we must learn the meticulous care required to connect things in unbroken succession, or rather, to derive things step by step. Even where we do not venture to apply mathematics we must always work as though we had to satisfy the strictest of geometricians.

In the mathematical method we find an approach which, by its deliberate and pure nature, instantly exposes every leap in an assertion. Actually, its proofs merely state in a detailed way that what is presented as connected was already there in each of the parts and as a consecutive whole, that it has been reviewed in its entirety and found to be correct and irrefutable under all circumstances. Thus its demonstrations are always more exposition, recapitulation, than argument. Having made this distinction, I may now return to something mentioned earlier.

We can see the great difference between a mathematical demonstration which traces the basic elements through their many points of connection, and the proof offered in the arguments of a clever speaker. Although arguments may deal with utterly separate matters, wit and imagination can group them around a single point to create a surprising semblance of right and wrong, true and false. It is likewise possible to support a hypothesis or theory by arranging individual experiments like arguments and offering proofs which bedazzle us to some degree.

But those who wish to be honest with themselves and others will try by careful development of individual experiments to evolve empirical evidence of the higher sort. These pieces of evidence may be expressed in concise axioms and set side by side, and as more of them emerge they may be ordered and related. Like mathematical axioms they will remain unshakable either singly or as a whole. Anyone may examine and test the elements, the many individual experiments, which constitute this higher sort of evidence; it will be easy to judge whether we can express these many components in a general axiom, for nothing here is arbitrary.

The other method which tries to prove assertions by using isolated experiments like arguments often reaches its conclusions furtively or leaves them completely in doubt. Once sequential evidence of the higher sort is assembled, however, our intellect, imagination and wit can work upon it as they will; no harm will be done, and, indeed, a useful purpose will be served. We cannot exercise enough care, diligence, strictness, even pedantry, in collecting basic empirical evidence; here we labour for the world and the future. But these materials must be ordered and shown in sequence, not arranged in some hypothetical way nor made to serve the dictates of some system. Everyone will then be free to connect them in his own way, to form them into a whole which brings some measure of delight and comfort to the human mind. This approach keeps separate what must be kept separate; it enables us to increase the body of evidence much more quickly and cleanly than the method which forces us to cast aside later experiments like bricks brought to a finished building.

The views and examples of the best men give me reason to hope that this is the right path, and I trust my explanation will satisfy those of my friends who ask from time to time what I am really seeking to accomplish with my optical experiments. My intention is to collect all the empirical evidence in this area, do every experiment myself, and develop the experiments in their most manifold variations so that they become easy to reproduce and more accessible. I will then attempt to establish the axioms in which the empirical evidence of a higher nature can be expressed, and see if these can be subsumed under still higher principles. If imagination and wit sometimes run impatiently ahead on this path, the method itself will fix the bounds to which they must return.[6]

April 28, 1792

A Bibliographical Essay on Goethean Approaches to Colour Science

The investigations of natural and experimental colour phenomena presented in this book are the result of a research tradition stretching back to Goethe's own investigations of colour. To make the current presentation of results from this research tradition as clear as possible, its historical dimension has, for the most part, been omitted from the main presentation. The purpose of this bibliographical essay is to provide an overview of this research and mention important contributions made over the last two centuries since the publication of Goethe's *Farbenlehre*. It is not possible here to give a comprehensive account of this particular history of science and many important researchers have inevitably been left out. What is presented here is primarily the research that directly influenced this book. For a comprehensive overview of the Goethean research tradition see Frederick Amrine (1987).

Amrine, Frederick. 1987. 'Goethe and the Sciences: An Annotated Bibliography'. In *Goethe and the Sciences: A Reappraisal,* edited by Frederick Amrine, Francis J. Zucker and Harvey Wheeler, 389–437. Dordrecht: Reidel.

The phenomena of nature

No book, not even all books taken together, can give an adequate account of the plethora of phenomena that arise and vanish in the ever-changing processes of nature. Although many of the fleeting colour appearances escape our hold on them, we can still find an impressive range of art installations that integrate natural phenomena into the realm of art, such as Andy Goldsworthy (1985; 1989). Painters such as J.M.W. Turner and Claude Monet perhaps came closest to capturing a passing moment of nature and its atmosphere. Their works are nicely presented by Michael Bockemühl (2015) and Daniel Wildenstein (2014) respectively. Bockemühl also addresses the influence of Goethe on Turner.

In nature photography we cannot find the same intensity as in Romantic and Impressionist painting, but nevertheless we can marvel at the wonderful pictures by Marcel Minnaert in his classic (1993) book. Two authors who particularly emphasise atmospheric phenomena are Robert Greenler (1999) and Michael Vollmer (2019). In addition, comprehensive accounts of the conditions under which atmospheric phenomena arise from the perspective of mathematical optics can be found in Bohren (2003) and Walker (2006). Johannes Kühl (2016) provides an account of atmospheric colour phenomena from a Goethean perspective. With the example of atmospheric colours, Kühl illustrates Goethe's experimental approach and the transition from natural phenomena to experimental phenomena.

Bockemühl, Michael. 2015. *Turner: 1775–1851; The World of Light and Colour*. Cologne: Taschen.

Bohren, Craig F. 2003. 'Atmospheric Optics'. In *The Optics Encyclopedia: Basic Foundations and Practical Applications,* edited by Thomas G. Brown, Katherine Creath, Herwig Hogelnik, Michael Kriss, Joanna Schmit, Marvin J. Weber, 53–91. Vol. 1. Berlin: Wiley-VCH.

Goldsworthy, Andy. 1985. *Rain, Sun, Snow, Hail, Mist, Calm*. Leeds: Henry Moore Centre for the Study of Sculpture.

—. 1989. *Leaves.* London: Art Data.

Greenler, Robert. 1999. *Rainbows, Halos, and Glories*. Milwaukee: Peanut Butter.

Kühl, Johannes. 2016. *Rainbows, Halos, Dawn and Dusk: The Appearance of Color in the Atmosphere and Goethe's Theory of Colors*. Hillsdale, New York: Adonis Press.

Minnaert, Marcel. 1993. *Light and Color in the Outdoors*. New York: Springer.

Vollmer, Michael. 2019. *Atmosphärische Optik für Einsteiger: Lichtspiele in der Luft*. Berlin: Springer Spektrum.

Walker, Jearl. 2006. *The Flying Circus of Physics*. Hoboken, New Jersey: Wiley.

Wildenstein, Daniel. 2014. *Monet: The Triumph of Impressionism*. Cologne: Taschen.

From natural phenomena to experimental phenomena

With his distinction between 'optics with the eye' and 'optics without the eye', Johannes Grebe-Ellis (2005) aptly characterises the difference between optics developed out of the observed phenomena and optics formulated within the paradigm of contemporary physics. A selection of contemporary research in phenomenological optics is also presented in Grebe-Ellis and Theilmann (2006). More than half of the books in the series Phänomenologie in der Naturwissenschaft edited by Grebe-Ellis and Lutz-Helmut Schön, of which the aforementioned conference proceedings was the first volume, deal with optical phenomena. This could well be because Goethe paid special attention to this field; the three-part *Farbenlehre*, first published in 1810, is by far his most comprehensive scientific work. There are many German editions of the *Farbenlehre*, the most popular is in the so-called Hamburger edition (Goethe 1981). This edition, however, does not contain the polemical part of Goethe's *Farbenlehre*. Douglas Miller translated the didactic part of the *Farbenlehre*, together with several other of Goethe's key scientific texts, including the essay 'The Experiment As Mediator between Object and Subject' included in this book. Twenty years before publishing the *Farbenlehre*, Goethe published a series of colour experiments, his *Beiträge zur Optik* (*Contributions to Optics*), which present a systematic approach to prismatic colour phenomena in unprecedented completeness. The original German is in Goethe (1951), a complete translation is in Goethe (1946), and translated excerpts are in Goethe (1971).

Other areas of optical phenomena, such as shadows, perspective and illumination, have been approached in a Goethean manner in by Georg Maier (2011). This book presents the first investigation of the inversion of shadows and the discovery of the 'light shadow' (see also Maier's articles on this topic in 2004). Maier's observational approach to these phenomena makes his work particularly useful for teachers. Draft curricula can be found in Mackensen and Ohlendorf (1998). This volume also develops quantitative relationships, such as the inverse-square law of light intensity in optics, without recurrence to assumptions based on physical models (see also Grebe-Ellis and Ohlendorf, 2011).

Many areas of optics have been scientifically investigated using a Goethean methodology. The field of polarisation is explored by Grebe-Ellis (2005). Refraction phenomena are dealt with by Wilfried Sommer (2005) and by Marc Müller and Lutz-Helmut Schön (2011). Thomas Quick (2015) has investigated refraction and its expression in the concept of elevation. His work was inspired by, among others, the unpublished research of Günther Altehage. Using the concept of optical elevation, Müller provides a phenomenological investigation of the rainbow (Grebe-Ellis and Müller, 2007). A detailed investigation of the transformation of an image seen through a glass sphere is given by Sascha Grusche, Matthias Rang and Marc Müller (2018).

The research mentioned so far deals with experimental approaches to colour phenomena without explicitly including properties of the eye. Following Goethe's approach of including the observer as part of the experiment, they do indeed make use of the eye as an instrument for analysing an optical configuration, but the properties of the eye itself and the resulting physiological phenomena are largely eschewed due to the focus on optical phenomena. With a method inspired by Goethe, the English colour researcher Michael H. Wilson and his colleague Ralph W. Brocklebank (1955) were able to demonstrate a difference between complementary pairs of colours produced by additive colour mixing and complementary pairs produced by afterimages. Wilson and Brocklebank (1960) also succeeded in showing that the sensational experiments with two colour projections presented by Edwin H. Land (1959) are examples of coloured shadows, a phenomenon first systematically investigated by Goethe. By developing an experimental series they not only placed Land's experiments into their proper scientific context, but also developed a better understanding of these phenomena. These papers, along with his other writings on colour are reprinted in Wilson (2018). With a similar methodological focus, Sebastian Hümbert-Schnurr (2018) quantitatively investigated physiological phenomena and their relation to optical phenomena.

The research presented in this section demonstrates that a Goethean approach can indeed lead to mathematical description. But this specialisation means that the literature is aimed primarily at scientists. Johannes Kühl (1988) outlines how some of these branches of optics – especially prismatic colours, turbidity colours and diffraction colours – can themselves be considered as a developmental series.

Goethe, Johann Wolfgang von. 1946. 'Contribution to Optics'. In *Pure Colour: Part 3; Extracts from Goethe's Scientific Work*. Translated by Eleanor C. Merry, 3–42. London: New Cultural Productions.

—. 1951. *Beiträge zur Optik und Anfänge der Farbenlehre: 1900–1808*. Edited by Rupprecht Matthaei. Weimar: Hermann Böhlaus Nachfolger.

—. 1971. *Goethe's Colour Theory*. Arranged and edited by Rupprecht Matthaei. London: Studio Vista.

—. 1981. *Werke. Hamburger Ausgabe in 14 Bänden.* 14 vols. Munich: C. H. Beck.

—. 1988. *Scientific Studies*. Edited and translated by Douglas Miller. New York: Suhrkamp.

—. 2005. *Grundzüge einer Phänomenologie der Polarisation*. Berlin: Logos Verlag.

Grebe-Ellis, Johannes, and Mark Müller. 2007. 'Spiegelbilder der Sonne im Tropfen – Zur Phänomenologie des Regenbogens'. *Didaktik der Physik*. Regensburg: Deutsche Physikalische Gesellschaft.

Grebe-Ellis, J. and Heinz-Christian Ohlendorf. 2011. *Lampensatz für Photometrieversuche*. Kassel: Bildungswerk Beruf und Umwelt.

Grebe-Ellis, Johannes and Florian Theilmann, eds. 2006. *Open Eyes 2005: Ansätze und Perspektiven der phänomenologischen Optik*. Berlin: Logos Verlag.

Grusche, Sascha, Matthias Rang and Marc Müller. 2018. 'Wie wird die Ansicht durch eine Kugellinse verformt? – Entwicklung einer Phänomenreihe'. *PhyDid B – Didaktik der Physik – Beiträge zur DPG-Frühjahrstagung*. http://phydid.physik.fu-berlin.de/index.php/phydid-b/article/view/890/1024.

Hümbert-Schnurr, Sebastian. 2018. *Farbe im Spannungsfeld zwischen Wahrnehmung und Messung: Entwurf eines phänomenologischen Zugangs zu Farberscheinungen*. Berlin: Logos Verlag.

Kühl, Johannes. 1988. 'Zum Goethe'schen Urphänomen der Farbentstehung und zu einem Zusammenhang mit Beugung und Brechung'. *Elemente der Naturwissenschaft* 49: 85–95.

Land, Edwin H. 1959. 'Experiments in Color Vision'. *Scientific American* 200, no. 5: 84–99.

Mackensen, Manfred von and Heinz-Christian Ohlendorf. 1998. *Modellfreie Optik: Grundlegende Themen aus dem Gesamtgebiet der Optik, mit Hinweisen zur Gestaltung von Physikunterricht der 12. Klasse*. Kassel: Bildungswerk Beruf und Umwelt.

Maier, Georg. 2004. *Blicken, Sehen, Schauen: Beiträge zur Physik als Erscheinungswissenschaft*. Edited by Johannes Grebe-Ellis. Dürnau: Kooperative Dürnau.

—. 2011. *An Optics of Visual Experience*. Hillsdale, New York: Adonis Press.

Müller, Marc and Lutz-Helmut Schön. 2011. 'Virtuelle Beugungsbilder am Gitter'. *PhyDid B – Didaktik der Physik – Beiträge zur DPG-Frühjahrstagung*. http://phydid.physik.fu-berlin.de/index.php/phydid-b/article/view/288.

Quick, Thomas. 2015. *Phänomenologie der optischen Hebung*. Berlin: Logos Verlag.

Sommer, Wilfred. 2005. *Zur phänomenologischen Beschreibung der Beugung im Konzept optischer Wege*. Berlin: Logos Verlag.

Wilson, Michael H. 2018. *What Is Colour? The Collected Works*. Edited by Laura Liska and Troy Vine. Berlin: Logos Verlag.

Wilson, Michael H., and Ralph W. Brocklebank. 1955. 'Complementary Hues of After-Images'. *Journal of the Optical Society of America* 45, no. 4: 293–299. Reprinted in Wilson, *What is Colour?*, 57–75.

—. 1960. 'Two-Colour Projection Phenomena'. *Journal of Photographic Science* 8: 141–150. Reprinted in Wilson, *What is Colour?*, 99–120.

Goethe's colour experiments
and their relation to physics

An unusual characteristic of Goethe's approach to colour is that he searched for the complementarity of physiological colours in the realm of physical phenomena. Surprisingly, he was able to extend the idea of complementarity from physiological to physical colours in the didactic part of his *Farbenlehre*. Rupprecht Matthaei (1971) provides valuable commentary on this endeavour. Goethe's search for general laws of colour expressed in the realms of both physiology and physics was not in accordance with the prevalent dualistic worldviews of his time. As a result, his *Farbenlehre* was largely rejected by physicists, although there were notable exceptions, such as Goethe's contemporary Johann Wilhelm Ritter, who discovered ultraviolet light based on Goethe's idea of complementarity. Olaf Müller provides an in-depth biographical account of this connection in (2021b) and an English summary in (2018).

Also problematic for the reception of Goethe's *Farbenlehre* was his rejection of claims made by the English physicist and mathematician Isaac Newton in his *Opticks*, first published in 1704, which in Goethe's time was regarded as the standard textbook on optics. With few exceptions, Newton describes experiments with a ray of sunlight in a darkened room, and his theory of colour reduces colours to components of this ray. Alongside the more mathematically ambitious, posthumously published *Optical Lectures*, the *Opticks* is Newton's opus magnum, which was published 35 years after his ground-breaking paper 'New Theory about Light and Colors'.

Goethe was one of the first scientists to realise that by limiting optical experiments to those done in darkness, Newton had drastically limited his empirical data. As a result, Newton's theory cannot be proven with his data; it can only be made plausible. This restricted approach resulted in the far-reaching complementarity of prismatic phenomena being systematically overlooked. Goethe published a critique of Newton's restricted experimental approach in the polemical part of his *Farbenlehre*, translated into English by Michael Duck and Michael Petry (2016). Attempts to refute Goethe's argumentation have been repeatedly made since its first publication, with Michael Duck's introduction to the English edition being but one example. Dennis Sepper (1988) provides a detailed assessment of the philosophical merit of Goethe's arguments. Evan Thompson develops Sepper's reconstruction of Goethe's critique of Newton into a general critique of contemporary colour science in (1995; see particularly Chapter 1).

Goethe's research programme for investigating the complementarity of optical colour phenomena was not developed for over a century. This was due, at least in part, to a lack of understanding of Goethe's argumentation. The psychologist August Kirschmann (1917; 1924) made ground-breaking discoveries regarding the importance of complementarity. He discovered complementary line spectra, demonstrated the suitability of both spectra for spectrometry, and discovered that spectral complementarity is not limited to the visible spectrum but extends into the infrared and ultraviolet regions. In a comprehensive work first published in 1946, P. J. Bouma (1971) provides the first colorimetric description of Goethe's edge spectra and shows that the CIE diagram (which is based on Newton's spectrum) could be based on the edge spectra instead. He also demonstrates that the Newtonian spectrum can be considered as arising from a subtractive mixing of edge spectra.

The importance of Kirschmann's pioneering work was not recognised for some decades. It was taken up in German-speaking regions by Max Barth (1944) and by Fritz Lobeck (1954). In Norway, Kirschman's research was developed by Andre Bjerke (1961). Bjerke was joined by, among others, physicist Torger Holtsmark (2012). In England, Bouma's research was taken up by Michael Wilson (1958). Wilson was joined by Ralph Brocklebank and they published a series of articles in leading physics journals (reprinted in Wilson 2018).

The Norwegian and English research groups worked simultaneously for a couple of decades developing Goethe's colour research in an academic context, without compromising on Goethe's original aims. While Wilson and Brocklebank have made important contributions especially to experiments on physiological colours, Bjerke and Holtsmark developed Goethe's prismatic experiments beyond Kirschmann's original insights. Holtsmark (1970) succeeded in proving theoretically that Newton's *experimentum crucis* can be both inverted and experimentally extended to include its complementary counterpart by using mirror apertures. He demonstrated that Newton's argument based on this experiment is not only methodically weak, but, as Abdelhamid Sabra (1981) has shown, also experimentally insufficient. The first person to experimentally realise Holtsmark's idea was the Swedish physicist Pehr Sällström, who contributed to the Scandinavian research impulse based in Oslo. This research is presented in the documentary film *Sällström* (2010).

Looking at the literature presented so far, despite a small selection we can see that before Wilson and Bjerke, experimental questions in particular had been dealt with almost exclusively by German-speaking researchers. With Wilson and Bjerke, two research centres emerged that had a major influence on subsequent research, including research in German-speaking regions. Olaf Müller gives an overview of this history in (2021a).

Since the turn of the millennium, the research tradition originating in Goethe's *Farbenlehre* has undergone further developments. The painter Ingo Nussbaumer showed that when coloured contrasts are used instead of the usual black and white ones, the complementarity of the spectra is preserved. Nussbaumer (2008) systematically documented for the first time the symmetry of these newly discovered spectra.

Continuing in the tradition of Holtsmark, Matthias Rang developed a mirror aperture that allows the simultaneous arising of complementary spectra from mutually dependent inverted images in Grebe-Ellis and Rang (2009). The mutual dependency of complementary spectra is shown to be a result of the conservation of energy in Grebe-Ellis and Rang (2018). Rang (2015) shows that a full inversion of complementary spectra – that is, an inversion not only of the image itself but also of its optical environment – leads to the concept of a 'lightened room', the optical counterpart to Newton's darkened room. In a lightened room Newton's *experimentum crucis* can be inverted, which is described by Olaf Müller and Rang (2009). Müller (2016) argues that every prismatic experiment necessarily has a complementary counterpart, and that it follows from his analysis of Goethe's experiments that Newton's theory of light is underdetermined and thus is only one possible explanation among others. Müller (2015) initiated an unusually active discussion of Goethe's *Farbenlehre* within academia, resulting in a special edition of *Journal for General Philosophy of Science* dedicated to the topic edited by Michael Hampe and Timm Lampert (2018). He also brings to the fore the strong influence of historical events on the reception of Goethe's works throughout its history. The edited volume Klug et al. (2021) is the result of Müller's emphasis on the role of experimental investigation for the reconstruction and clarification of experimental questions in the history of science.

A summary of recent experimental results on Goethe's optical experiments can be found in Grebe-Ellis and Passon (2020).

Barth, Max. 1944. *Beiträge zur experimentellen Erweiterung von Goethes Farbenlehre: Zum Studium von Goethes Farbenlehre 2*. Dornach: Versuchsraum für Goethes Farbenlehre.

Bjerke, Andre. 1961. *Nye bidrag til Goethes farvelære: Første del: Goethe kontra Newton*. Järna: Kosmos Förlag.

—. Forthcoming. *Contributions to Goethe's Theory of Colour*. Edited by Johannes Grebe-Ellis, Arne Nicolaisen, and Troy Vine. Berlin: Logos Verlag.

Bouma, P. J. 1971. *Physical Aspects of Colour: An Introduction to the Scientific Study of Colour Stimuli and Colour Sensations*. London: Macmillan.

Goethe, Johann Wolfgang von. 1971. *Goethe's Colour Theory*. Arranged and edited by Rupprecht Matthaei. London: Studio Vista.

—. 2016. *Goethe's 'Exposure of Newton's Theory': A Polemic on Newton's Theory Of Light and Colour*. Translated by Michael Duck and Michal Petry. London: Imperial College Press.

Grebe-Ellis, Johannes and Oliver Passon. 2020. 'Goethe's Farbenlehre from the Perspective of Modern Physics'. *Holistic Science Journal* 4: 50–59.

Grebe-Ellis, Johannes and Matthias Rang. 2009. 'Komplementäre Spektren: Experimente mit einer Spiegel-Spalt-Blende'. *Mathematisch Naturwissenschaftlicher Unterricht* 62, no. 4: 227–31.

—. 2018. 'Power Area Density in Inverse Spectra'. *Journal for General Philosophy of Science* 49: 515–523.

Hampe, Michael and Lampert, Timm, eds. 2018. 'Goethe and Newton on the Theory of Colours'. Special issue, *Journal for General Philosophy of Science* 49, no. 4.

Holtsmark, Torger. 1970. 'Newton's Experimentum Crucis Reconsidered'. *American Journal of Physics* 38, no. 10: 1229–1235. Reprinted in *Colour and Image*, 35–50.

—. 2012. *Colour and Image: Phenomenology of Visual Experience.* Edited by Johannes Grebe-Ellis. Berlin: Logos Verlag.

Kirschmann, August. 1917. 'Das umgekehrte Spektrum und seine Komplementärverhältnisse'. *Physikalische Zeitschrift* 18: 195–205.

—. 1924. 'Das umgekehrte Spektrum und die Spektralanalyse'. *Zeitschrift für Instrumentenkunde* 44: 173–175.

Klug, Anastasia, Olaf Müller, Anna Reinacher, Troy Vine, and Derya Yürüyen, eds. 2021. *Goethe, Ritter und die Polarität: Geschichte und Kontroversen.* Paderborn: Brill-Mentis.

Lobeck, Fritz. 1954. *Farben anders gesehen: Neue Ergebnisse zur Farbenlehre Goethes.* Basel: Die Pforte.

Müller, Olaf L. 2015. *Mehr Licht: Goethe mit Newton im Streit um die Farben.* Frankfurt am Main: Fischer Verlag.

—. 2016. 'Prismatic Equivalence – A New Case of Underdetermination: Goethe vs. Newton on the Prism Experiments'. *British Journal for the History of Philosophy* 24, no. 2: 323–347.

—. 2018. 'Goethe and Ritter'. In Vine, *Experience Colour*, 150–159.

—. 2021a. 'Hundert Jahre Dunkelheit? Nachwort: Zur Geschichte der polaristischen Forschung an Goethes *Farbenlehre*'. In Klug et al., *Goethe, Ritter und die Polarität*, 320–351.

—. 2021b. *Ultraviolett: Johann Wilhelm Ritters Werk und Goethes Beitrag – zur Biographie einer Kooperation.* Göttingen: Wallstein.

Müller, Olaf L., and Matthias Rang. 2009. 'Newton in Grönland: Das umgestülpte experimentum crucis in der Streulichtkammer'. *Philosophia Naturalis* 46, no. 1: 61–114.

Newton, Isaac. 1672. 'New Theory about Light and Colors', *Philosophical Transactions of the Royal Society* 80: 3075–87.

—. 1952. *Opticks, or, A Treatise of the Reflexions, Refractions, Inflexions and Colours of Light.* New York: Dover Publications.

—. 1984. *Optical Lectures: 1670–1672.* Edited by Alan E. Shapiro. Cambridge: Cambridge University Press.

Nussbaumer, Ingo. 2008. *Zur Farbenlehre: Entdeckung der unordentlichen Spektren.* Vienna: Splitter.

Rang, Matthias. 2015. *Phänomenologie komplementärer Spektren.* Berlin: Logos Verlag.

Sabra, Abdelhamid. 1981. *Theories of Light from Descartes to Newton.* Cambridge: Cambridge University Press.

Sällström, Pehr. 2010. *Monochromatic Shadow Rays: A Film about Experiments on the Rehabilitation of Darkness*. DVD. Stuttgart: Edition Waldorf.

Sepper, Dennis L. 1988. *Goethe contra Newton: Polemics and the Project for a New Science of Color*. Cambridge: Cambridge University Press.

Thompson, Evan. 1995. *Colour Vision: A Study in Cognitive Science and the Philosophy of Perception*. New York: Routledge.

Vine, Troy, ed. 2018. *Experience Colour: An Exhibition by Nora Löbe & Matthias Rang*. Nailsworth: The Field Centre.

Westphal, Jonathan. 1991. *Colour: A Philosophical Introduction*. Oxford: Blackwell.

Wilson, Michael H. 1958. 'Goethe's Colour Experiments'. *Year Book of the Physical Society* 12: 12–21. Reprinted in *What is Colour?*, 77–92.

—. 2018. *What Is Colour? The Collected Works*. Edited by Laura Liska and Troy Vine. Berlin: Logos Verlag.

Goethe's methodology

Goethe's investigation of the complementarity of colour phenomena presented in the *Farbenlehre* serves as an exemplar for Goethe's scientific approach to nature as a whole, which seeks to understand the essence of nature as the expression of polarities. However, Goethe only made passing remarks on this larger project, for example, in the 'Introduction to the Lectures on Physics' (1971, 68). In the didactic part of the *Farbenlehre*, Goethe mentions that his search for polarity in nature is based on a methodological analogy with the field of magnetism. Goethe's conception of polarity, however, is not static but dynamic. He develops a conception in which a polarity is reciprocally determined by its multiple phenomenal manifestations, a process which Goethe calls *Steigerung* (enhancement). This shows methodological connections between the *Farbenlehre* and Goethe's project of morphology, which is outlined in his essay 'Toward a General Comparative Theory' (Goethe 1988, 53–69) and applied in *The Metamorphosis of Plants* (Goethe 1988, 73–110).

Besides the idea of polarity and enhancement, of central importance to Goethe's approach to science is finding relations between phenomena through a process Goethe calls *Vermannigfaltigung* (manifolding). Goethe presents this process in his essay 'The Experiment As Mediator between Object and Subject', included in this book. Friedrich Steinle (2002) shows that it was not only Goethe who worked in this way but also scientists such as Faraday. Such approaches become especially common in the formative stages of a scientific paradigm. Steinle established the term 'exploratory experimentation' to distinguish this kind of experimentation from the theory-orientated experimentation that happens once a paradigm has been established (see also Ribe and Steinle 2002).

The aim of Goethe's methodology should not, however, be seen as merely providing descriptions of phenomena or a pre-theoretical treatment, as Gábor Zemplén (2001) shows in his doctoral thesis. Grebe-Ellis (2011), in his introduction to a new edition of optical works by Rasmus Bartholin and Christiaan Huygen, shows that not only Bartholin but also Newton at times experimented in a manner similar to Goethe. Such approaches to science were thus established scientific practices in Goethe's time (and still are). Indeed, as Nesbit (1972) shows, Goethe's methodology draws deeply on the scientific tradition developed by Francis Bacon and Robert Boyle, which Thomas Kuhn refers to as the 'experimental tradition' in his classic (1977) essay.

As a result, aspects of Goethe's method can be found in both contemporary science and the early modern experimental tradition, to which Newton also belonged. A central idea of the experimental tradition is that experimentation should be based on an analysis that proceeds from phenomena to the principles underlying the phenomena. Troy Vine (2020) shows that Goethe's approach is not a rejection of this analytical method and, moreover, characterising Goethe's method as synthetic in opposition to the analytic methods of experimental science is a misrepresentation of Goethe's own understanding of his method and its relation to science. Indeed, as Vine shows in (2021), Newton's and Goethe's shared analytical method is the basis of Goethe's critique of Newton.

Goethe's recognition of the observer as an integral part of an experiment made his methodology particularly appealing to pioneers of quantum physics. Goethe's approach received praise from physicists such as Werner Heisenberg (1958) and Carl-Friedrich von Weizsäcker (1987).

Max Born (1963) dedicated his speech at the twelfth meeting of Nobel Prize laureates in Lindau to Goethe's *Farbenlehre*, acknowledging Michael Wilson's work as an important continuation to understanding Goethe's *Farbenlehre*.

Another aspect that appealed to these pioneers of quantum physics was Goethe's rejection of the mechanical worldview that had been adopted by early modern science and dominated scientific thought in Goethe's time. The American physicist Arthur Zajonc (1998) shows how quantum physics challenges this mechanical worldview. In this paper Zajonc also shows that Goethe's approach is in this respect even more radical, which is probably why Heisenberg, Weizsäcker and Born ultimately considered it incompatible with physics. English quantum physicist Henri Bortoft (1996), on the other hand, shows that the wholeness demonstrated by quantum physics is the same kind of wholeness as the wholeness disclosed by Goethe's method. Bortoft develops this common conception of wholeness into an influential interpretation of Goethean science. Bortoft's work has also influenced contemporary interpretations of quantum physics, such as that developed by the mathematician Louis H. Kauffman (2020).

While the pioneers of quantum physics found Goethe's methodology ultimately too radical, it was most congenial to the German Idealists searching for a new scientific approach to nature in the wake of Immanuel Kant's critical philosophy. Philosopher Eckart Förster (2012) reconstructs this philosophical development and the central role played by Goethe's methodology in the philosophical systems of Schelling and Hegel. This philosophical reconstruction also provides historical support for Bortoft's interpretation of Goethe based on the conception of wholeness in quantum physics. Rudolf Steiner also placed Goethe in the context of German Idealism in his introductions in the so-called Kürschner Edition of Goethe's scientific works (reprinted in Goethe 1982). While writing these introductions (translated in 2000), Steiner developed a fuller philosophical interpretation (translated in 2008).

Like Hegel before him, Ludwig Wittgenstein was disillusioned with contemporary scientific attempts at understanding colour and looked to Goethe for inspiration. In his *Remarks on Colour* (1979), he applies Goethe's methodology to an investigation of the logic of colour. Drawing on Wittgenstein's insights, Jonathan Westphal (1991) develops a Goethean investigation of the complementarity of the absorption spectra and colour of objects using Michael Wilson's colour research. Affinities between Goethe's scientific method and Wittgenstein's philosophical method are explored by Joachim Schulte (1990) and Mark Rowe (1991). Marc Müller (2017) explores parallels between Wittgenstein's and Goethe's method and applies them to physics.

What appealed to Steiner and Wittgenstein alike is that Goethe's methodology provides a way of investigating nature that is based on seeing the world differently. By replacing mechanical explanation with a richer, fuller perceptual experience of phenomena, the phenomena can explain themselves. This idea is developed by Vine (2018) in the context of Wittgenstein's philosophy. This addresses the aspect of Goethe's approach to science that Frederick Amrine aptly calls 'the metamorphosis of the scientist' in (1988). This transformational aspect of Goethe's methodology is developed in the context of Goethe's *Farbenlehre* by Johannes Kühl and Matthias Rang (2020); in the context of the history of physics by Arthur Zajonc (1993), and in the context of continental philosophy by Henri Bortoft (2012). The full relevance of Goethe's methodology for the evolution of consciousness is made explicit by Owen Barfield (1988).

Amrine, Frederick. 1988. 'The Metamorphosis of the Scientist'. In Seamon and Zajonc, *Goethe's Way of Science*, 33–54.

Barfield, Owen. 1988. *Saving the Appearances: A Study in Idolatry.* Middletown, CT: Wesleyan University Press.

Born, Max. 1963. 'Betrachtungen zur Farbenlehre'. *Die Naturwissenschaften* 50: 29–39.

Bortoft, Henri. 1996. *The Wholeness of Nature: Goethe's Way of Science.* Edinburgh: Floris Books.

—. 2012. *Taking Appearance Seriously: The Dynamic Way of Seeing in Goethe and European Thought.* Edinburgh: Floris Books.

Förster, Eckart. 2012. *The Twenty-Five Years of Philosophy: A Systematic Reconstruction.* Translated by Brady Bowman. Cambridge, MA: Harvard University Press.

Goethe, Johann Wolfgang von. 1982. *Naturwissenschaftliche Schriften*. Edited with introductions, footnotes and commentaries by Rudolf Steiner. 5 vols. Dornach: Rudolf Steiner Verlag.

—, 1971. *Goethe's Colour Theory*. Arranged and edited by Rupprecht Matthaei. London: Studio Vista.

—. 1988. *Scientific Studies*. Edited and translated by Douglas Miller. New York: Suhrkamp.

Grebe-Ellis, Johannes. 2011. 'Bild und Strahl: Perspektiven der Optik bei Bartholinus und Huygens'. Introduction to *Versuche mit dem doppeltbrechenden isländischen Kristall, Abhandlung über das Licht*, by Erasmus Bartholinus and Christian Huygens, vii–xliv. Frankfurt: Harri Deutsch.

Heisenberg, Werner. 1958. 'Die Goethesche und die Newtonsche Farbenlehre im Lichte der modernen Physik'. In *Wandlungen in den Grundlagen der Naturwissenschaft*, 85–106. Stuttgart: Hirzel.

Kauffman, Louis H. 2020. 'Compresence and Coalescence'. *Holistic Science Journal* 4: 24–39.

Kühl, Johannes, and Matthias Rang. 2020. 'A Model for Scientific Research: A Consideration of Goethe's Approach to Colour Science'. *Holistic Science Journal* 4: 60–71.

Kuhn, Thomas. 1977. 'Mathematical versus Experimental Traditions in the Development of Physical Science'. In *The Essential Tension: Selected Studies in Scientific Tradition and Change*, 31–65. Chicago: University of Chicago Press.

Müller, Marc. 2017. *Grammatik der Natur: Von Wittgenstein Naturphänomene verstehen lernen*. Berlin: Logos Verlag.

Nesbit, H. B. 1972. *Goethe and the Scientific Tradition*. London: University of London.

Ribe, Neil M. and Friedrich Steinle. 2002. 'Exploratory Experimentation: Goethe, Land, and Color Theory'. *Physics Today* 55, no. 7: 43–49.

Rowe, Mark. 1991. 'Goethe and Wittgenstein', *Philosophy* 66, 283–303.

Schulte, Joachim. 1990. 'Chor und Gesetz: Zur "morphologischen Methode" bei Goethe und Wittgenstein'. In *Chor und Gesetz: Wittgenstein im Kontext*, 11–42. Frankfurt am Main: Suhrkamp.

Seamon, David, and Arthur Zajonc. 1998. *Goethe's Way of Science: A Phenomenology of Nature*. Albany: State University of New York Press.

Steiner, Rudolf. 2000. *Nature's Open Secret: Introductions to Goethe's Scientific Writings*. Translated by John Barns and Mado Spiegler. New York: Anthroposophic Press.

—. 2008. *Goethe's Theory of Knowledge: An Outline of the Epistemology of His Worldview*. Translate by Peter Clemm. Great Barrington, MA: Steiner-Books.

Steinle, Friedrich. 2002. 'Experiments in History and Philosophy of Science'. *Perspectives on Science* 10, no. 4: 408–432.

Vine, Troy. 2018. 'Polemics as Therapy'. In *Experience Colour: An Exhibition by Nora Löbe & Matthias Rang,* edited by Troy Vine, 168–77. Nailsworth: The Field Centre.

—. 2020. 'Newton, Goethe and the Mathematical Style of Thinking: A Critique of Henri Bortoft's *Taking Appearances Seriously*'. *Holistic Science Journal* 4: 72–87.

—. 2021. 'Goethes Newton-Kritik als interne Kritik'. In *Goethe, Ritter und die Polarität: Geschichte und Kontroversen*, edited by Anastasia Klug, Olaf Müller, Anna Reinacher, Troy Vine, and Derya Yürüyen, 31–58. Paderborn: Brill-Mentis.

Weizsäcker, Carl-Friedrich von. 1987. 'Goethe and Modern Science'. In *Goethe and the Sciences: A Reappraisal*, edited by Frederick Amrine, Francis J. Zucker and Harvey Wheeler, 115–132. Dordrecht: Reidel.

Westphal, Jonathan. 1991. *Colour: A Philosophical Introduction*. Oxford: Blackwell.

Wittgenstein, Ludwig. 1979. *Remarks on Colour*. Edited by G. E. M. Anscombe. Oxford: Blackwell.

Zajonc, Arthur. 1993. *Catching the Light: The Entwined History of Light and Mind*. Oxford: Oxford University Press.

—. 1998. 'Light and Cognition: Goethean Studies as a Science of the Future'. In Seamon and Zajonc, *Goethe's Way of Science,* 299–134.

Zemplén, Gábor A. 2001. 'An Eye for Optical Theory? Newtons Rejection of the Modificationist Tradition and Goethe's Modificationist Critique of Newton'. PhD thesis, Budapest University of Technology and Economics.

Acknowledgements

This book started out as the exhibition catalogue *Experiment Farbe*, published in 2010. It was written by Nora Löbe and Matthias Rang together with Johannes Kühl for experiments they designed for an exhibition of colour phenomena at the Goetheanum in Switzerland. The exhibition was designed by scientists of the Science Section at the Goetheanum in collaboration with artists to celebrate the bicentennial of Goethe's *Farbenlehre*, which was first published in 1810.

In 2018 the exhibition was presented in Glasshouse College, a Ruskin Mill College in Stourbridge. The catalogue *Experience Colour* was produced for this exhibition, and was an extensive English revision of the original German catalogue undertaken by Nora, Matthias and Troy Vine. The present book is, in turn, an extensive revision of the first half of the English exhibition catalogue.

It is in the nature of a project of this sort to incur many debts of gratitude along the way. We would like to thank everyone who helped us. In particular, we would like to thank Arthur Zajonc for writing the Foreword, Sonja Dorau for translating the original German exhibition catalogue and a draft of the bibliographical essay and Johannes Grebe-Ellis and Leo Meyer for comments on the manuscript. We would like to thank the team at Floris Books, in particular Christian Maclean for suggesting this publication, David Budd for his indefatigable effort in making the text as clear as possible, and Jenny Skivington for the exquisite layout.

Our special thanks go to Johannes Kühl, then leader of the Science Section, for making the original exhibition in Switzerland possible, and Aonghus Gordon, founder of Ruskin Mill Trust, for bringing the exhibition to England and supporting the development of new exhibits. This book is dedicated to them.

Picture Credits

Public domain: Figures 1.1, 4.12, 5.1, 6.1, 6.9, 7.1, 7.2, 7.3, 7.5, 7.8, 7.9; p. 132.

Böttge, Susanne: Figures 1.2, 3.6, 3.9, 3.20, 3.26, 3.48 (bottom).
Clarke, Gordon: Figure 4.1.
Craig, Keith: Figures 1.8, 1.10, 1.11, 2.1, 2.8, 3.4, 3.7, 5.7, 5.18, 6.6, 7.6 (bottom).
Jäggi, Hugo: Figures 3.12, 3.13, 3.38.
Kühl, Johannes: Figures 3.4, 3.5.
Löbe, Nora: Figures 1.3, 1.4, 1.5, 1.6, 1.7, 2.2, 2.3, 2.4, 2.6, 2.7, 2.9, 2.11, 2.12, 2.13, 3.1, 3.2, 3.3, 3.11, 3.14, 3.15, 3.16, 3.17, 3.18, 3.27, 3.28, 3.31, 3.32, 3.33, 3.34, 3.35, 3.36, 3.39, 3.40, 3.41, 3.42, 3.43, 3.45, 3.46, 3.47, 3.50, 3.51, 3.52, 3.53, 3.54, 3.55, 3.56, 3.57, 3.60, 3.61, 4.2, 4.3, 4.4, 4.5, 4.6, 4.7, 4.8, 4.9, 4.11, 4.16, 4.22, 4.23, 5.8, 6.2, 6.3, 6.4, 6.5, 6.7, 6.8, 6.10, 6.11, 6.12, 6.13, 6.14, 7.4, 7.6 (top), 7.7; pp. 10, 142.
Onneken, Johannes: Figures 2.14, 2.15, 2.16.
Rang, Matthias: Cover images; figures 1.9, 2.5, 2.6, 2.9, 2.10, 3.8, 3.12, 3.19, 3.21, 3.22, 3.23, 3.24, 3.25, 3.29, 3.30, 3.37, 3.44, 3.48 (top), 3.49, 3.58, 3.59, 4.13, 4.14, 4.15, 4.17, 4.18, 4.20, 4.21, 5.2, 5.3, 5.4, 5.5, 5.6, 5.9, 5.10, 5.11, 5.12, 5.13, 5.14, 5.15, 5.16, 5.17, 6.12, 6.13, 6.14; pp. 4, 128.
Reakes, Simon: Figure 7.10.
Sojka, Wladyslaw (www.sojka.photo): Figure 4.19.
Vine, Troy: Figures 3.10, 4.10; pp. 6, 9, 160, 164.

Endnotes

1. Goethe, Johann Wolfgang von. 1999. *Werke: Hamburger-Ausgabe in 14 Bänden*, vol. 14, *Naturwissenschaftliche Schriften II*. Munich: dtv, 259. (We have provided our own translation as there is no English translation of the third, historical part of the *Farbenlehre*.)
2. Goethe, *Werke*, 14: 97–98 (our translation).
3. Kuhn, Thomas. S. 2012. *The Structure of Scientific Revolutions*. Chicago: University of Chicago Press, 119.
4. Goethe, Johann Wolfgang von. 1988. *Scientific Studies*. Edited and translated by Douglas Miller. New York: Suhrkamp, 181.
5. Goethe, *Scientific Studies*, 307.
6. Goethe, *Scientific Studies*, 11–17.

Index

You may also be interested in...

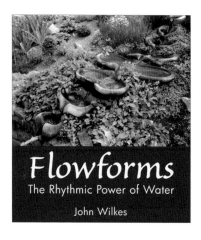

Flowforms
The Rhythmic Power of Water
John Wilkes

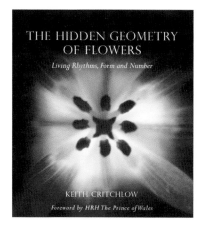

THE HIDDEN GEOMETRY OF FLOWERS
Living Rhythms, Form and Number

KEITH CRITCHLOW

Foreword by HRH The Prince of Wales

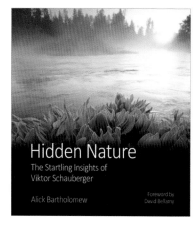

Hidden Nature
The Startling Insights of Viktor Schauberger

Alick Bartholomew

Foreword by David Bellamy

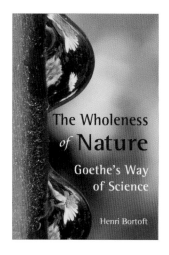

The Wholeness of Nature
Goethe's Way of Science
Henri Bortoft

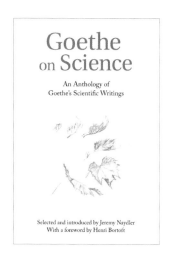

Goethe on Science
An Anthology of Goethe's Scientific Writings

Selected and introduced by Jeremy Naydler
With a foreword by Henri Bortoft

Taking Appearance Seriously
The Dynamic Way of Seeing in Goethe and European Thought

Henri Bortoft

Floris Books

For news on all the latest books, and to get
exclusive discounts, join our mailing list at:

florisbooks.co.uk/mail/

And get a FREE book
with every online order!

We will never pass your details to anyone else.